Washed Up

Also by Skye Moody

Fiction

The Good Diamond

Medusa

K Falls

Habitat

Wildcrafters

Blue Poppy

Rain Dance

Nonfiction

(as Kathy Kahn)

Fruits of Our Labor

Hillbilly Women

WASHED UP

The Curious Journeys of Flotsam & Jetsam

SKYE MOODY

SASQUATCH BOOKS
SEATTLE

In memory of my mother,
Donna Kelly,
who taught me to float.

Printed in the United States of America
Published by Sasquatch Books
Distributed by Publishers Group West
15 14 13 12 11 10 09 08 07 06   9 8 7 6 5 4 3 2 1

Cover and interior design/composition: Stewart A. Williams
Interior photographs: Skye Moody, except p. 49 (Gary Luke) and p. 203 (Vince DeWitt/*Cape Cod Times*)

Cover photograph ©2003 by David Liittschwager and Susan Middleton, all rights reserved. From the book *Archipelago: Portraits of Life from the World's Most Remote Island Sanctuary* by David Liittschwager and Susan Middleton, National Geographic Books, 2005. Photograph was taken on Laysan Island in the Northwestern Hawaiian Islands.
Author photograph: Rosanne Olson ©2006

Library of Congress Cataloging-in-Publication Data is available.

ISBN 1-57061-463-6

Sasquatch Books
119 South Main Street, Suite 400
Seattle, WA 98104
(206) 467-4300
www.sasquatchbooks.com
custserv@sasquatchbooks.com

Shall I part my hair behind? Do I dare to eat a peach?
I shall wear white flannel trousers, and walk upon the beach.

—T. S. Eliot, "The Love Song of J. Alfred Prufrock"

Flotsam: *n. 1) Wreckage of a ship or its cargo found floating on the surface of the sea. Usually associated with* **jetsam**. *2) Timber, etc., washed down by a stream.*

Jetsam: *n. Goods discarded from a ship and washed ashore; material thrown overboard in order to lighten a vessel (also called waveson). Usually associated with* **flotsam**.

Lagan: *n. Also* **lig- (lagmar=dragnet)**. *Goods or wreckage lying on the bottom of the sea.*

Contents

Ocean Surface Currents © 2006 Dr. Michael Pidwirny

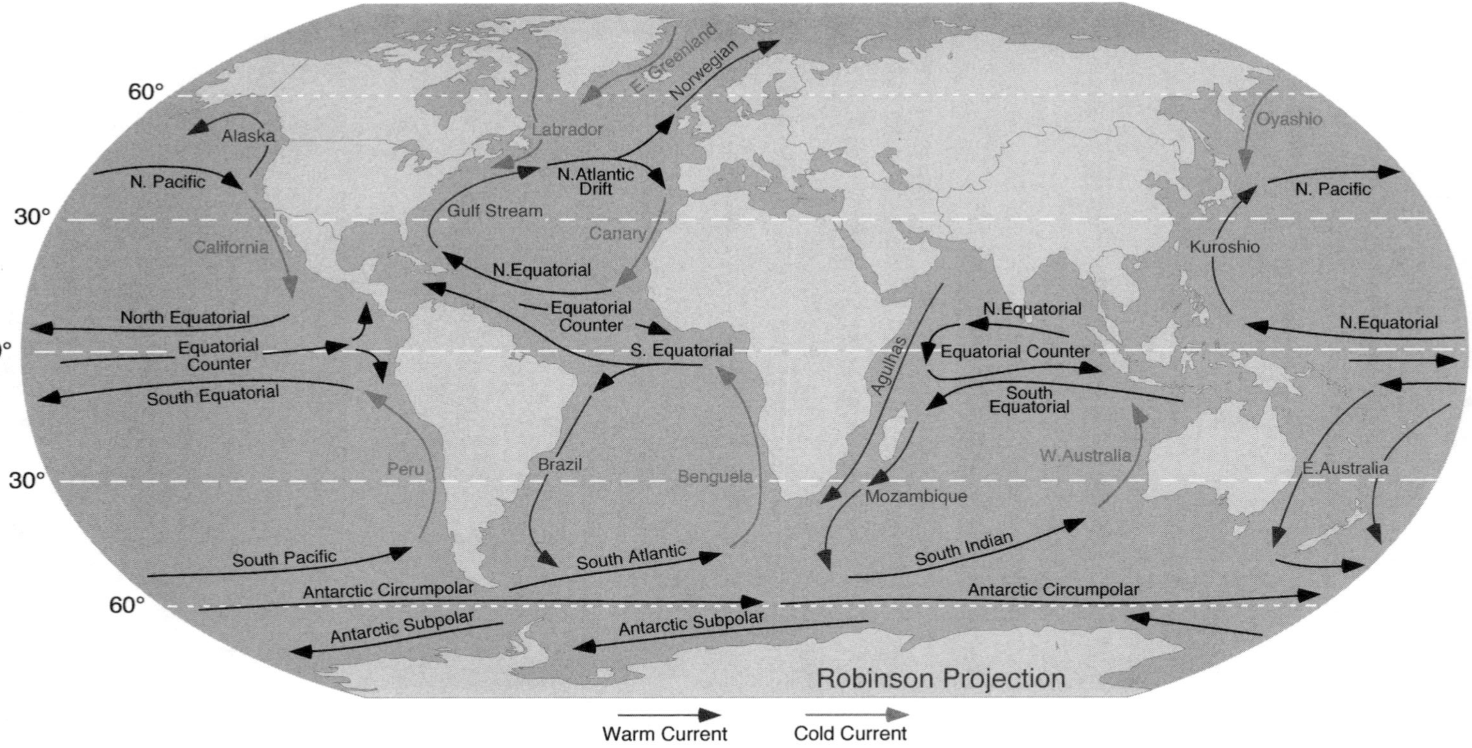

TIDE LINES: *An Introduction*

The Nautical Gods Must Be Crazy

Can a stone float? I plucked it from a rock pile six feet above the tide line. Holding it in the palm of my hand, the apricot-colored stone felt too light for a rock. It was the size of a hen's egg, or a peach pit on steroids. Its narrowest end was tipped black as if dipped in an inkwell, and pockmarks riddled its surface, the way brick looks after it has lain in saltwater for a while. Yet when I ran a fingernail across its orange surface, it didn't powder like brick or tint my fingernail orange. I looked around. Why, out of millions of rocks that wash up on this beach, a varietal bazaar of colors, shapes, textures, and striations, had I singled out this particular stone? What had drawn my fickle flotsamist's eye to a stone I could easily discard as just another boring chunk of baked clay washed up on the tides?

For one thing, the last high tide had shoveled ashore this pile of stones hours ago, and now beneath a burning noonday sun, the multicolored mountain of rocks had dried to a dull gray, as colorful stones will when coated with saltwater. This orange stone lay on top of the heap. It had not dulled when dried, and it stood out the way an agate, when dry and salt-coated, still emits a luminescence that separates it from duller opaque stones. This stone wasn't glowing like an agate, but its vivid color popped despite saltwater and the sun's heat.

I carried the stone to the water's edge and placed it on the becalmed ebbing tide. The greater the density of an object in relation to the amount of water it displaces, the deeper it sinks to the bottom. Spheres, like this stone, can be deceptive; the stone seemed too light to sink, yet too heavy to float. I pressed it between thumb and forefinger. It didn't react to the pressure. I held it against my ear and shook it, but couldn't tell if it was hollow.

Yet it floated. So, flotsam. That it remained floating on the water's surface suggested an air pocket inside. Ridiculous, I sneered, and snatched it up off the tide.

Turning it over in my hand, I had to make a decision: My pockets, even the hood of my sweatshirt, were chockablock with flotsam I had scored that morning along a lonely stretch of Alki Beach on the shores of Puget Sound in Washington State. I had no room for another object. If I kept this peculiar stone, I'd have to toss out the little pink plastic propeller, or the white coral, or the Jagermeister screwtop with the Exacto knife inside, or the driftwood knot shaped like a talking duck's head. No, I couldn't afford to sacrifice any of my treasures for a mutant peach pit.

Years ago, in Tokyo, I accompanied a Japanese friend to a Shinto temple. The quarter-mile path to the shrine was graveled with tiny gray stones. After posting our petitions, as we returned along the gravel path, I suddenly had an inexplicable urge to lean over and pick up a certain piece of gravel, a specific stone indistinguishable from the millions of others, except it seemed to scream "Pick me up!" Stooping, I picked up the screamer, about the size of a nickel. Holding the tiny stone in my palm, I turned it over. A human face had been drawn on it.

Why, out of several million gravel rocks, had I chosen the one with a face drawn on the other side? The other possibility was that every stone on the Shinto path had a face drawn on its other side. For years afterward, that image haunted my dreams.

A lifetime of flotsam collecting has invested a certain discernment into my beachcombing practices. I've learned to distinguish unusual flotsam from ordinary flotsam, and I'm very picky about what I collect from the world's beaches. This morning at Alki Point, everything in my pockets and hood had been judiciously selected as a keeper. In the end, I photographed the peculiar stone and left it on the beach near the tide line where I'd found it.

Six hours later, the highest tide of the year washed over the beach. At ebb, its powerful wet fingers clawed and raked and rearranged the rocks and logs, leaving a new wrack line and a completely rearranged beach. Anything that floated had caught the seaward drift, including the mysterious stone.

You'd think after the Japanese gravel experience I would have learned my lesson. I still have that tiny piece of Shinto gravel with the face drawn on it. Alas, like the fool on his journey, I failed to look back into my knapsack of experience, and so years later, I let the special beach stone wash out on the tide.

That evening I studied the picture I'd taken of the stone that floated. Now beyond my grasp, it seemed even more compelling. Something uneasy in—yes—the pit of my stomach told me I'd made a terrible mistake. After all, I could have carried the stone home in my mouth.

A stone light enough and impermeable enough to float. Or had I simply romanticized a chunk of brick that had been burned on one end? But wait. Check out this cool pink plastic propeller. Where do you suppose it floated in from? I decided it came off a Japanese boy's toy airplane and had floated across the Pacific Ocean to wash up on a beach in Puget Sound. Now that's a great flotsam story.

Flotsam Nightmares

The nightmares changed. Now instead of a million gravel faces, I dreamed oceans of ovoid bricks came floating toward shore. As they washed onto the beach, they self-constructed a Great Wall of America all along the Pacific Coast, an impenetrable barrier whose single purpose was to strand my piscine nature from my spawn mates out at sea. On the beach side, I hammered and chiseled at the growing brick wall; more bricks floated in and reinforced the wall. Then, just before waking in sweat-soaked sheets, I saw the strange stone floating out of reach on the water side of the brick wall, a sacred object mocking my foreshortened graspy fingers.

I have tossed floating objects into the Sound only to find them, several days later, back on the beach in approximately the same spot, give or take a few yards, where I'd thrown them, their traveling speed and distance dependent on the winds and the water's surface currents. I've even experimented with the phenomenon by tossing unusual flotsam, even rocks, into the water, finding them washed up nearby again the next day, and the next, and the next.

A week after photographing the stone and enduring its taunting nightmares, I returned to the beach in search of it. Starting at the lighthouse, at approximately the same spot where I'd found the stone, I worked methodically along the tide line. If the Coast Guard was watching, they surely marveled over the hooded figure digging madly through heaps of beach rocks in the pouring rain. I'm lucky they didn't report me to Homeland Security.

It was there somewhere, I could feel it. I dug like a Labrador pawing for a bone. I'd make one hole, fill it back in and dig another, and so on, until the stones themselves began complaining: "What are you? Some kind of cargo cultist who's mislaid your talisman?"

They were right. In one week I had gone from a simple, devoted collector of tidal gifts to a full-fledged, raving cargo cultist, believing in my heart and soul that this particular stone I'd found and spurned was actually a gift from the nautical gods meant for me alone, my talisman, my own little maritime miracle.

I looked into cargo cults. The more I read, the more I knew I was one with them, a worshiper of exotic phenomena inexplicably materializing in my own private marine world. Only the gods could have sent this mystery flotsam. Like all novice cargo cultists, at first I thought the gods must be crazy. Venerate a floating stone? "Flotsam worshiper!" hissed my inner ear demons. Come on, get a life.

I printed several copies of the photograph, placing the stone's image all over the house: on the refrigerator in a magnet frame, in the laundry closet where my husband never went, in my private loo where he wouldn't see how obsessed I'd become. Again, I reminded myself this was just flotsam. This thing I had so foolishly abandoned yet could not forget had been reclaimed by the sea, and only the gods knew where it was headed now—certainly to someone more worthy. I followed currents up and down the beach, scanning with binoculars. It couldn't have gone far. Or could it? Days passed, then weeks. I must have pawed through a thousand rock piles, having long ago given up filling in the holes, instead leaving little volcanolike structures as I moved methodically from one section of beach to another.

Other beachcombers began grumbling about my disruption of the strandline's natural pattern. I began to resent their intrusion into my search. They were only interested in pretty seashells, fully intact—heaven forbid they'd collect a cracked or otherwise deformed oyster shell—or beach glass, preferably violet hued—God, the fights I've seen over violet-hued beach glass. They're the type who snub their noses at the really cool beach finds, walk right past a Japanese flip-flop sandal, or a rusted sign for A&W Root Beer, or a pair of men's Hanes jockey briefs washed up in a bright navy-blue swirl among the wrack. No, these beachcombers with their little plastic drugstore bags take no interest in barnacle-coated soccer balls, itinerant geoduck harvesting licenses, rusted oil drums, mysterious light bulbs, and little plastic toys. Litter, they label it in their pitiable ignorance. If they resented my hole digging, I sniffed at their competitive head butting over trivial bits of colored glass. Talk about cargo cultists.

For half a millisecond I considered asking one beachcomber if he'd seen my lost stone, and then realized he wouldn't have noticed it if it had hit him between the eyes; he had no more interest in floating rocks than I had in perfect oyster shells, the latter an object so redundant on this tide line you might as well name it Beach Rockefeller. Anyway, Oyster Shell Man was probably collecting shells for his garden. A lot of people do that; they come to the beach and harvest oyster shells to crush into bone meal and feed to their roses. I don't know if this practice is legal, or even if it offends the ecological sensibilities of nondiggers. I wonder, though, if Neptune snickers at such minor scrapes on the fringe of his domain.

Weeks of searching the strandline and pawing through stone heaps failed to produce my precious flotsam stone. I made another copy of the photograph for my car and pasted it to the dashboard. I did not yet resort to petitioning the saint of things lost. St. Anthony is a last resort and, besides, I still had a few cards up my sleeve before calling up the Patron Ace of Carelessness. I vowed to reverse my sin of omission, my failure to accept with grace the gift Neptune had offered me to have and to hold, to keep and to worship. No soul-stealing photograph could replace the real thing. I would find it, or I would find its match, and no measure of grumbling about disturbing the natural strandline would deter me. And by the way, flotsam worship is not confined to egg-shaped stones. In fact, the little pink plastic propeller I plucked from the sand enjoys a prime position on my flotsam altar. This is no ordinary gift from the sea. After all, how many pink plastic propellers have the oceans coughed up?

By the Wind Blown

In a sense, we are all flotsam and jetsam. Created in watery media, we wash up without personal choice on life's foreign shores, only to navigate unpredictable currents and tides, blown from one circumstance to another, until we land, finally, for the last time, eventually bereft of flesh and humors, a mere carcass whose tactile memory of its adventures has been only partly etched into its hollowed bones. Like so much flotsam, what's left of us tells some, but not all, of the story. Like flotsam, our personal journeys will never be fully understood or explained. And thank the gods for that, for the journey is who we are; without our personal mystery and its individual transformations and permutations, we'd be as indistinguishable as grains of sand viewed by the naked eye, our legacy unthinkably transparent.

Flotsam of Flotsam

This much I know about the saltwater in my veins: In 1898, Master Mariner Captain Alfred J. Green, for thirty-five years commander of British sailing ships, saw the inevitable demise of the magnificent masted barques and clippers, those tall ships, those seagoing gems, being rapidly replaced by steam-powered vessels. Captain Green saw the future and it wasn't made of canvas sheets. Bowing to modern technology, which he knew included the dumbing down of all aspects of sailing, from relationships among crew members to the formal traditions and pristine conditions of these vessels, Captain Green didn't exactly abandon the sea; he decided to view it from a rocky beach. That year, with his wife and six children, he departed England's port city of Liverpool aboard a steamer and crossed the Atlantic to Montreal, Canada. The family traveled by train to British Columbia, eventually sailing southward into Puget Sound, by then officially located in the United States of America. Captain Green had visited this part of the world during his many sea voyages and had determined that no place on the globe could match the natural beauty of the U.S. Northwest. Rather quickly the family settled on ten acres, including six hundred feet of shoreline, on Vashon-Maury Island at Quartermaster Harbor. They had a house built and named the family settlement Stillhaven, perhaps because this marked the location on the globe where the Green family set aside navigational charts as stillness replaced moving with the tides.

Captain Green was my great-grandfather, the fifth generation of English sea captains who skippered British sailing ships. My grandfather Alan was just a boy of six when the family settled on this American beach, but he had already traveled the world aboard the British barque *Wilhelm Tell*, his father at the helm. In his book *Jottings from a Cruise*, a collection of his father's ships' logs and a personal memoir, Alan recalled that Stillhaven had "a nice gravel beach, where Indian arrows were frequently found, and where every tide brought in an interesting assortment of flotsam, the greatest portion being driftwood."

On that same island, more than fifty years later, I too beachcombed, and there, my feet in Keds to protect them from barnacles, I learned to swim. Bouyant saltwater lends a sense of security; surely a body will float to the surface every time. I could lie on my back, arms and legs outstretched, and float, it seemed, forever, until something distracted me and I disturbed the balance between my body and the water it displaced. If I didn't start swimming, I'd sink. I was proud of my ability to float, bobbing on the waves, drifting with the currents. I suppose this is when I first recognized my flotsam heritage, and first became fascinated with the odd bits of flotsam and jetsam washing up on the beach.

On idyllic summer days, I'd awake to the low burr of the foghorn at Point Robinson lighthouse. I couldn't wait to hit the tide line, and I ran without caution along a nettled path leading to the beach because I knew some great new discovery washed up with each new tide. We found life rings on the beach, pieces of boats, discarded boating supplies, whiskey bottles, fishermen's nets, cork floats, barnacle-coated glass fishing floats from Japan, and once my brother, Rob, found a live octopus, weighing around twenty pounds, beached on sun-toasted rocks. He fetched a wheelbarrow, plopped the octopus into it, and wheeled it up a steep hill to show our grandmother; then he rolled the wheelbarrow back to the beach and set the octopus afloat. I've often wondered what the octopus thought about his brief landlubbing adventure.

My older sister, Suzie, was enamored of sand-frosted bottles and frosty glass shards that washed up. She collected her sea jewels in a clear glass jar, submerging them in freshwater to show them to best advantage. Some white glass pieces that had over time developed a purple cast were her favorites even before she learned the explanation for their distinctive violet hue.

The Romans collected beach glass, as did other ancient Mediterranean cultures. Tumbled by the action of water and sand, broken glass over time washes up on beaches, transformed into uniquely frosted translucent shards. Artists have found innumerable uses for beach glass, from mosaics to jewelry to leaded glass windows and lamps, picture frames, mobiles, and more abstract creations. Probably the most desirable sea glass, or beach glass, is the lavender glass my sister preferred. It originated in the United States during World War I, when U.S. manufacturers were unable to figure out the correct chemicals for making a type of glass they had previously purchased from Germany. The glass version they did manufacture, based on a slightly altered formula, turned lavender, or light violet, with age. This purple or lavender glass is easily distinguishable and can be dated. Because glass is biodegradable, its presence in the ocean does not generally harm the marine environment. I suppose a fish might swallow a sharp piece of glass to disastrous effect. Beyond that, glass returns eventually to its original form: sand.

More plastic than glass now washes up along the world's beaches; still, glass shards etched over time by contact with underwater rocks and sand wash up on beaches everywhere. Fully intact glass bottles, these days, are indeed a rare treasure. Beach glass has become so scarce that glassmakers now manufacture a product called beach glass, meant to substitute for the real thing. Glass flotsamists shun the fakes.

Rocks, their shapes and striations, fascinated me in my early flotsaming days. Warmed dry by the sun or still cold and coated with seaweed slime, the rocks on my childhood beach were of many colors, some bright red or green, some translucent yellow, some sparkling with fool's gold or silver, and others simply smooth gray, green, or black spheres. At four years old, I had no idea that many of these rocks had originated high in the Olympic and Cascade mountains, had tumbled downstream on fresh mountain rivers that poured into the ocean, and had then been shoveled ashore by powerful incoming tides. The best rocks went into my sunsuit's pockets, and I filled them until their weight forced me to hunch over and almost crawl—still, each was a keeper. Imagine my disappointment when, once I'd hauled them up the hill to my grandparents' summer house and set them on the porch to dry, most of the keepers turned dull and uninteresting; shape shifters of the mineral world. If they didn't pass the dry test, I'd toss them back into the Sound, and if they skipped, I counted them worthwhile.

Riding incoming tides, seashells got tossed up to form a delicate white necklace on the beach, each shell also a keeper. Once again my sunsuit pockets filled until I could barely walk. Each shell had a story. Some still housed tenants, and these also went immediately back into the Sound, a reprieve from death by beaching, at least in my imagination. The best shells were lined with mother-of-pearl—oyster shells or deep blue mussels—or were clam shells larger than my father's hand, bleached chalky white, deep ridged semi-spheres, best when discovered with joints intact, two shells forming a nautical castanet.

And bones—fish bones, whale bones, seal bones, maybe even human bones—the remains of living creatures thrilled a four-year-old's soul. I created a life-and-death history for each bone I collected off the tide. A pirate's finger. A sea lion's claw. Once, in Canada, on a low tide off Point Roberts, I found an entire horse's skull. Sand crabs had moved in.

Having traveled the world on Great Britain's last great tall ship, my grandfather taught us to respect everything that came from the sea. This included the masses of driftwood flotsam that washed ashore to decorate the high tide line along the beach. Huge logs of old growth trees, uprooted or sawed down, made barefoot walking less daunting than on the rocky beach, if you didn't count splinters. When high tide hit and the waves chased us up the beach, we'd scramble onto the highest logs, timing our escape just right for maximum risk and thrill. Some logs still had root systems and branches intact; standing on end as they had in the ancient evergreen forests, they would have dwarfed the nearby lighthouse. Driftwood came in many shapes and sizes, often bleached and weathered to a fine silver sheen. Every day new logs washed up, some so gnarled they looked like sea monsters rising from the deep.

Whenever my grandfather walked on the beach, he would have, besides his pipe and his panama hat, a walking stick—always chosen from a driftwood pile at the beginning of his walk and always returned to the pile at walk's end. He had learned as a young boy aboard ship that everything must have a purpose; it was either ballast or had some other practical application, with a few exceptions for aesthetic necessities—the piano in the captain's quarters, for example. But everything that came from the sea must be returned to the sea; everything that the tide tossed onto land must either be used for good purpose or placed back on the beach where it had washed ashore, until another wave raked it adrift once more.

Like my ancestors, I have spent nearly a lifetime traveling the world relentlessly, have washed up at many exotic ports and gone adrift again in search of new adventure. Throughout my travels I've collected flotsam, jetsam, and lagan from the world's beaches, often provoking startled reactions from customs officials. Why would anyone want five kilos of seashells from the Black Sea? What kind of contraband is an old Coke bottle with Russian labeling sealed with a note inside written in Chinese characters? And what are these tiny bones here in your cosmetic bag?

Yet even while I acknowledge my flotsam origins and lifestyle, constantly washing up and going adrift again, I had to ask myself, when all is said and done, am I simply a beachcomber? Or am I a true-blue flotsamist? The mysterious, befuddling floating stone had triggered this question, launching me on a quest to discover its origins, and the difference between a mere beachcomber and a flotsamist. This called for plunging into the very essence of flotsam and jetsam.

I. Flotsam's Noble Origins

Little by little, when the tide receded, we made our way down among the crags until we came to a strip of seashore, and from this point we could see that the island was of large size, its interior being sheltered from storms by the front of the mountain. But what took our wonder was this: on the seashore was amassed the wealth of a thousand wrecks. Scattered here, there, and everywhere, in foam and high and dry, were flotsam and jetsam of richest merchandise, much of it spoiled by the sea, but much more cast high up and still of great value.

SINBAD THE SAILOR
THE SIXTH VOYAGE
One Thousand and One Nights

Ancient Flotsam

Who knows what the earliest sailors jettisoned from their outriggers and canoes? Surely anything that impeded forward motion in an emergency was tossed overboard to sink, swim, or bob off on a passing current. In fact, before the concept of synthetics took hold, everything that went into the sea consisted of organic matter, and much of it floated. Surely ancient mariners jettisoned their spoiling dinner leftovers, emptied snuff spittoons, and so forth—offal to the gods. When sailors figured out that ballast made for a safer, steadier, and more balanced ride, they used rocks to weigh down their boats. Eventually rocks were replaced with lead, and then manufactured goods. In a pinch, when the ship's hull was hit by hostile gunfire and a hole resulted, or stormy weather threatened to sink the ship, some or all of the ballast was jettisoned in order to raise the hull high enough to keep the hole out of tempestuous seas. Under the most treacherous circumstances, they jettisoned precious cargo, chests of gold and jewels, even, I dare say, human cargo—which did not usually go willingly into the big drink.

Most ballast sank, but some, such as bloated corpses, if not gobbled up by scavengers, or snagged and entangled in Davy Jones's graspy fingers, eventually washed ashore. And so jetsam became flotsam for a time, until it beached—or until it sank and became lagan. Over the eons since Poseidon first thrust his furious trident into solid land, flotsam and jetsam have gained generic status in the vernacular, yet its origins are maritime in nature: Flotsam refers to articles found floating on or slightly beneath the ocean's surface, some of it washing ashore along the tide line. Flotsam can be human, arriving via the oceans in boats, or washing up as corpses drowned at sea. Or flotsam can be drifting wood, or Nike athletic shoes, refrigerators, television sets, rubber ducks, or any object the sea churns to the surface. What's important about flotsam is that, at least for a part of its life journey, it floats. Jetsam is whatever is jettisoned into the water, including ballast, cargo, plastic soda bottles, glass fishing net floats, bales of marijuana, soccer balls, hockey gloves, plastic ducks, or dead bodies. If it floats, jetsam then becomes flotsam. If it sinks to the ocean floor, it becomes lagan, such as the remains of ancient shipwrecks littering that watery grave mariners dread: Davy Jones's locker.

Little is known about man-made flotsam and jetsam before the era of Europe's tall ships, but the earliest beachcombers were at least as curious as today's strandliners and certainly paid heed to what washed up on their beaches. Eskimos dwelling on remote, treeless Arctic islands for centuries constructed their homes entirely from flotsam driftwood, some arriving from as far away as Polynesia. Kayaks, too, were fashioned from driftwood frames upon which sea otter, or seal, hide was stretched taut. One-hundred-percent organic ocean material went into the ancients' homes and conveyances. Coastal natives throughout the world depended upon this early form of Home Depot delivered via Federal Express—in this case, drifting currents—and so invested the oceans with a powerful "gifting" reputation, though the early gods of the sea weren't always in a benevolent mood: The sea could turn against a sailor, even against whole coastal communities, its ferocious, often hair-trigger temper attributed to powerful gods dwelling within the rogue waves blown out of proportion by Triton's fickle conch.

Ancient Greece identified Poseidon as the temperamental god ruling the watery realm, but then the Romans, who had to own everything, changed Poseidon's name to Neptune; still the same god with the same surly temper. This is not important, except to point out that a god by any name never deserts his realm.

Whether pirates' treasure, a ship's hull, or a human skeleton, lagan often finds a new purpose as aquatic shelters for sea creatures, at least until an enterprising diver discovers it and hauls the treasure up into a boat. Now it's loot. Most undersea treasure hunters let old bones alone, either in respect for the dead or because they aren't worth salvaging. Human bones don't bring much on the market. Some lagan rests buried beneath ocean sediments until Neptune pitches a tantrum and shoves his little finger into the soup, generating massive undercurrents that rip up sediment and send lagan afloat, tumbling into a shallower current, where the stuff might travel for centuries before sinking again, or breaking up, hurling treasures like pieces of eight tumbling ashore. Flotsam of the gods. Which, naturally, leads to cargo cults.

The concept of cargo cults was popularized in the film *The Gods Must be Crazy*, in which an isolated African tribe discovered a Coke bottle that fell from the sky, and believing it had come from the gods, treated it as an object of worship. What Karl Marx called "commodity fetishism" was called "cargo cults" by social anthropologists. Both labels described the nineteenth-century phenomenon of European explorers and missionaries along with their ships and cargo washing up on the shores of Melanesia. Melanesians believed the marvelous Western goods arriving on tall ships were created by their ancestors and sent from the sea, where white people hijacked what rightfully belonged to the Melanesian descendants. The natives reacted by performing rituals including imitating the behavior of the white people in the belief that this mimicry would attract more cargo from their ancestors. They constructed elaborate miniature artifacts of

European ships and their inhabitants (and after World War II, when cargo was dropped from aircraft via parachutes, they made miniature planes and airports), believing that new cargo would land on the replicas. They set the miniatures, like shrines, along the tide line and waited for treasures to wash up. And they did. Only problem was that white folk, with their nasty habits and queer religious rites, accompanied the goods.

Such respect for flotsam is not necessarily infused with religious or apocalyptic inferences. Yet, unlike the Melanesian cargo cults whose interpretation of Europeans washing up in huge boats epitomizes the flotsam fetish, Pacific Islanders believed the Europeans who made sudden appearances on their beaches were actually ghostly spirits of their ancestors returning from their ritual burial grounds at sea. Others thought the pale-skinned beings unloading the gods' wonderful cargo were at worst hijackers, at best merely the gods' delivery vehicle—not really humans, more like UPS delivery agents, aliens from the Brown Planet—who had no value beyond delivering the goods. So they killed them.

Flotsam not only changes lives, but it can also transform an entire community's world view. Imagine beachcombing on the island where you had lived all your life and finding a stranded crate containing a refrigerator. But you've never seen a refrigerator before, don't understand its purpose, or even why the sea gods sent it to you. But they did, and what the gods send is either good or evil, depending on which god dispatched the item. In your spiritual eschatology, all sea gods are venerable; therefore, what is delivered up from the sea is good, and meant to serve as food, shelter, clothing, tools, art, or even an object worthy of worship. In this case, the refrigerator isn't applicable to any of your day-to-day needs, and as you caress its smooth finish, its curves and its corners, you perceive it as more than aesthetically pleasing; indeed, you experience a heightened state of mind, a spiritual rush, and recognize that the sea gods have sent you a religious artifact meant to be worshipped. Welcome to the cargo cults.

God as Flotsam

As recently as August 2000, a millenarian, or apocalyptic, form of cargo cult sparked to life on a beach in Indonesia's Moluccas Islands. This one involved God as flotsam, and he seemed to have arrived on the last tide. In his first appearance, God suffered from *kasado*, a terrible skin condition, and performed humanly impossible feats such as piercing a coconut with a grass straw. In the second instance, when God washed up on the beach, he was apparently suffering from *latah*, a speech condition similar to Tourette's syndrome. Cursing, he chased a young boy, threatening him with a knife. Both cases of God as flotsam are hotly debated today, although the second washed-up God may have been a Bugis from Sulawesi. The Bugis tribe, native to Sulawesi, operate a commercial sailing fleet, are Muslim, and are religious and tribal enemies of the minority Christian Moluccans, who discovered the latah-stricken apparition on their beach.

Oceania in particular, because of its many islands and therefore many beaches, embraces what washes in from the seas as spiritually infused. Usually these beliefs are tied to powerful supernatural beings and to dead ancestors or rulers. Micronesian and Polynesian spiritual beliefs involve cults of gods and heroes. When early Europeans overlaid the natives' pagan mystical animism with Christianity, the two melded into a sort of polytheistic magical Jesus and Mary cult, all this being window dressing for the natives' deepest spiritual connection—with the gods of the sea. In Oceanic and Arctic societies, rituals exhort intervention of sea spirits to clear stormy weather, deliver whales, perk up the fishing and life in general. In the Coral Islands, natives believe two sea spirits, Soalal and Mar, capture the souls of the dying and take them to the bottom of the sea to live. Mar is a good spirit who communicates with the living. Soalal is evil and causes illness and sometimes misfortune at sea. When people died from accidents or childbirth, they were buried at sea to minimize their spiritual influence on land. Their bodies were wrapped in mats, weighted, and offered adrift to Soalal. Some sank; others bobbed and floated along on currents that carried their bloating, gaseous corpses to distant beaches where they were hailed by total strangers as the returning bodies of long-lost ancestors. They were prayed over, sometimes lavish parties were thrown for them, and once again they were set adrift until, at long last, they sank for good.

The lesson here is that proper weights are extremely important, as is knowledge of winds and currents. Every flotsamist knows this. You only want to launch a dead body once.

This reminds me of a wake I once attended in Finland. It was winter solstice. Snow carpeted the ground. We'd been drinking Finlandia since 11 a.m. in a narrow bar on a Helsinki mews just off a small public park. We were holding a wake for the departed summer. My friend Alpo Suhonen was at the time coach of the Finnish national hockey team. On the third or fourth round, Alpo brought up the subject of burial at sea. This led to my remark on the cultural value of flotsam. The very subject of flotsam so electrified Alpo that he nearly levitated off the bar stool, and we passed two forgettable hours reminiscing over what we've plucked off ocean beaches.

Around 3 p.m., as Alpo gradually disintegrated into his glass, Anttii, a grandson of Sibelius, joined us. Anttii is a filmmaker, so gorgeous he wears permanent clutch marks from the women he constantly fends off. Anttii, too, was enthralled with the subject of flotsam. And so the wake for summer turned into a flotsam fest, each of us swapping our best flotsam stories, much like fishermen swap tales. It was nearing midnight when we were startled out of our pseudo-philosophical doldrums by a certain Finnish rock star, who shall go without a name because I've forgotten it. A tall drink of water, livelier on his feet than most Finns in winter, he announced that he was shortly departing for an African tour. Alpo mumbled something sacastic like, "I hope you get malaria." The rock star slid up to my side. "All the snow got you down?" he asked.

"Not at all," I said. "There's something very spiritually cleansing about snow." I was drunk.

The rock star sucked his cheeks. "Oh yeah? Why don't we go outside and you can show me what's so holy about snow."

Never one to reject a challenge, especially after a few drinks, I second-lined out into the dark mews where a single lantern cast our shadows across the small park. I scoured the scene for a clean patch of snow. "Over there," I said. "I'll make an angel appear in the snow."

String Bean howled. "You remind me of that Afro-Brazilian touching cult," he said. "You know the one. They believe if you touch something a spirit will appear."

He was confusing the Umbandan flotsam cults with the Bugis from Sulawesi, but I didn't contradict him. I lay down on my back in the snow. Spread my arms and legs. I'd show him how to make an angel appear. I flapped my limbs. And then I started sinking. Two, three inches at first, and as the ice cracked and separated, my feather down coat and sheepskin boots took in water and suddenly, I was floating—yes, floating on water.

"Oh God," the rock star cried. "You are—how do you say it?"

"Drowning . . ."

But no. I floated, until a gentle wave stirred up by breaking ice pushed me ashore. There is, after all, a god of flotsam in city ponds.

Within the Arctic Circle, thirteen distinct groups known collectively as the Inuit, scattered across Russia, Alaska, northern Canada, and Greenland, have for five thousand years worshiped sea spirits, as well they should, living in such a split-personality climate. I don't know why anyone would want to live four months a year in total darkness with plunging temperatures. And I suspect if the ancient Inuit had known about Hawaii, they would have risked death on grueling seas to get there. Forget whale blubber and reindeer aortal blood; anyone can live on coconuts and sunshine. As it was, the Inuit depended upon their honored spirits for the strength to endure such harsh conditions. Animism played a role in their belief systems: The offer of a drink to a dead seal pleased its spirit. Drink, because the seal's soul lived in its bladder; the drink would expedite the soul's return to the seas from which most of the Inuits' material possessions came. An important Inuit goddess was Sedna, who lived at the bottom of the sea and controlled whales, seals, and other sea life critical to survival. Coastal Native Americans, too, worshipped sea spirits and believed they sent the things that washed up on their beaches. Everything that washed up from the oceans was flotsam from the gods. But I'll bet that didn't include hula-hula girls.

Tides, currents, and waves are as old as the oceans they travel. Riding them, all manner of organic materials have ventured forth to foreign lands. Tree limbs from Oceania have traveled the Pacific Ocean's encircling currents some eleven thousand miles to beach in North America and Canada. According to legend and the Vancouver Maritime Museum, the earliest recorded human-made flotsam, a carved totem pole, washed up on North America's Pacific Coast several thousand years ago, stranding on a beach in Canada's Queen Charlotte Islands. Cultural anthropologists cite carbon dating and carving patterns as evidence that the Polynesian log inspired the Haidas' totem-carving culture, arguably an early form of trademark infringement.

In any case, the first Native American beachcomber to stumble upon that carved Polynesian log must surely have been awed and perhaps spiritually stirred, if not terrified, by the strange and wonderful thing.

Even today some beachcombers find personally relevant signs or mystical messages in the whorls of a seashell, or in the shape of a piece of driftwood, or even in the swirling of seaweed in the intertidal zone; like reading tea leaves, seaweed patterns speak to them; they go home filled with a sense of having received new insights into their lives and their *raison d'etre*. I recently met a woman in Garibaldi, Oregon, who was traveling America's coastlines on her personal "vision quest" in a belief that the ocean would deliver a critical message providing her with spiritual guidance, said message arriving on the tides and intended specifically for her.

My encounter with the vision-questing lady occurred near the ocean shore, inside a small shop crammed behind hundreds of gnarled and silvery lengths of driftwood flotsam. The shop owner handcrafted birdhouses and sold them, along with driftwood. I had noticed the shop owner's eyes held a special glint, which at first I attributed to smoking a blunt. Then the vision-questing lady floated into the shop, her eyes equally twinkling, her countenance aglow. She fished a small object from her pocket and held it clasped in her fisted hand. As the shop owner's eyes met the woman's eyes, the electricity got too woo-woo for me, so I ducked outside to make pictures of the driftwood.

Moments later the vision-questing lady swept out of the shop and into her vehicle, a dull, dust-sheathed car, and peeled rubber down Highway 101. Then the shop owner came out to show me what the woman had given him: a tiny crystal pyramid. It seems she had been on the beach near his shop the day before and had found some personal insight or spiritual message washed up especially for her. During her receipt of spiritual flotsam, she had encountered the birdhouse man, whom she immediately linked to her tide-line epiphany. And so the following day she tracked him down at his shop to present her new god with this little crystal offering.

Moments this charged with woo-woo occur all over the world, all the time, yet they seem to occur with frightening frequency in coastal communities where all those negative ions flow in off the spindrift and heighten the senses. By now the vision-questing lady may have reached the Florida Keys. You will know her by her dull, forgettable car, and the unbearable lightness of her gait.

I have often wondered what America's earliest beachcombers thought when they first tripped over a dead pilgrim washed up on their beach, or a one-hundred-liter jug of Spanish wine impossibly intact lodged between beached driftwood. Imagine, if you will, the French fisherman whose boat drifted into the wrong current, at long last sighting land and bringing his tired vessel ashore only to be met by terrified and perhaps indignant natives. Thus was born Acadia. But even before the newcomers sloshed ashore, beachcombers along the world's coastlines were discovering amazing natural flotsam; some they used for food, some for decorative jewelry, some for medicine, and some for construction of homes and boats. The abundance of native references to whaling and other nautical pastimes, and to weather conditions over the ocean, indicate the natives, however they interpreted their experiences, knew about shifting currents, tides, and winds, often taking advantage of a particular current to travel to fishing grounds or to explore potential campsites. Early natives intuitively understood the nature of the great ocean and the winds that blew across it, and depicted its varied temperaments as god spirits. Thus the god Thunderbird, for example, might travel a current's path straight to shore, embracing in its waves amazing gifts from the sea. Thunderbird beached whales for the sake of the Inuit. Thunderbird brought fallen trees for building shelter, and seals for making oil and clothing, great rafts of seaweed for food and medicine. (Personally, I'll take the powder-blue convertible with whitewall tires.)

Returning to the cargo cult phenomenon: Among philosophical archaeologists—this hot new subspecialty has recently chipped off from archaeology's Mycenaean roots—the tired joke about cargo cults is that all coastal natives want a refrigerator to wash up on their beach. The improbable event actually occurred on September 21, 2004, along the Pembrokeshire coast of Wales, where several dozen refrigerators crashed onto the rocks at Precipe Bay. Alas, the rocks smashed the refrigerators to smithereens. It's believed a storm in the Bristol Channel ejected the cargo off the MV *Ryfgell* as it sailed from Dublin bound for Avonmouth. Besides refrigerators, the accidentally jettisoned cargo included some crates of raisins. The raisins washed up at Caldey Island where beachcombers from nearby Tenby reported "sea grapes" appearing on their beaches.

Dragon Spittle Fragrance

> *All along the shore were planks and fragments of many vessels that had been wrecked on this inhospitable coast. And this was not all, for when we proceeded through the island, we found a spring of pure ambergris overflowing into the sea; and by this the whales are attracted, but when they have swallowed it and dived to the depths of the sea it turns in their stomachs and they eject it, so that it rises to the surface in solid lumps such as are found by sailors. But the ambergris that is cast about the opening of the spring melts in the heat of the sun, and its perfume is blown about the island, wafted sweet upon the breeze like fragrant musk.*
>
> SINBAD THE SAILOR
> THE SIXTH VOYAGE
> *One Thousand and One Nights*

When Sinbad discovered ambergris flowing from a spring into the sea, Scheherazade's sailor almost got it right. Sinbad describes ambergris as a substance that attracts whales, who ingest the stuff and then, finding it disagrees with them, eject it into the ocean, where the malodorous material "rises to the surface in solid lumps."

In Africa, ambergris was traded as far back as 1000 BC, the mysterious substance coveted for its fragrance. Still, no one seemed quite certain exactly what the substance was, only that it came from the ocean and floated on its surface, sometimes beaching along the African coastline, at other times harvested straight from the sea in great lumps the size of a rich man's hut. In Egypt, ambergris was burned as incense, but its origin remained unknown. The Japanese also possessed and traded in ambergris, calling it *kunsurano fuu*, or "whale feces," indicating they had at least partly figured out where it came from. But ancient India held that ambergris was derived from whale vomit, coming closest to the true explanation. The Chinese had the most inventive theory on ambergris, calling it *lung sien hiang*, meaning "dragon spittle fragrance," because it was believed that the waxy substance drooled into the ocean from the mouths of dragons sleeping on sea rocks. Dragon drool is still sold in Chinese apothecaries as an aphrodisiac, as incense, and as a spice.

The baffling properties of ambergris must have seemed like the ancient version of the magic bullet. Arabs used it to treat diseases of the brain and heart. The Arab trader Ibn Haukal in the tenth century lauded its aphrodisiac quality, suggesting its worth was roughly equal to the price of gold, or the price of African slaves. Meanwhile, the Greeks thought ambergris enhanced the soothing effects of wine, often slipping a little into their guests' cups. Turkish Muslims carried ambergris along on their pilgrimages to Mecca as the ultimate offering to Allah. During Iran's Satiric period, Goshtasb "ordered that fires should be lighted and myrrh and ambergris be ignited." Persians ate ambergris along with hashish "without guilt," according to the chronicler Ibn Battuta.

In *The Travels of Marco Polo*, the thirteenth-century explorer refers to the collecting of ambergris on island beaches off the coast of India and Madagascar. By then ambergris enjoyed the distinction of being the world's most prized and desired flotsam.

Perfumers, herbalists, and shamans through the centuries have claimed that the scent of properly aged ambergris directly interacts with human sexual appetites (pheromones) and have often touted the substance as an aphrodisiac. Europeans once generally agreed that bees living in hives near the seashore produced ambergris, but some argued that it was a form of seabird or sperm whale guano. Louis XV reputedly flavored his favorite meals with ambergris, and Queen Elizabeth I used ambergris to perfume her gloves. In Elizabethan times, ambergris rivaled gold in value.

The word ambergris derives from the Arabic *anbar*, for a kind of tree, which may have produced resin. Also owing its name to this Arabic word is amber jaune, the yellow fossil resin of 40-million-year-old trees, found along Baltic seashores and prized as jewelry even in ancient times. The equivalent to the word anbar in Greek is *elektron*, the root of the word electricity. When liquefied into a tincture, ambergris gives off the same warm and energetic light that the fossilized resin—amber—radiates. Interestingly, though fossilized amber and ambergris come from two entirely different sources, both are found on beaches, both float in seawater, and both appear rather innocuous until human hands get hold of them. Still, no one could fully explain the existence of ambergris until some startled sailor apparently witnessed an actual event at sea, and eventually the mystery was solved.

Imagine a seasick Moby Dick.

Ambergris is a waxy substance related to cholesterol, produced in the lower intestinal tract of the sperm whale, forming around accumulated indigestible matter such as octopus remains, cuttlefish bones, and squid beaks. When too many squid beaks and cuttle bones accumulate in a sperm whale's digestive tract, a curious chemical process takes place in which the irritating refuse forms into a waxy blob. Eventually the whale's stomach rumbles and it belches—or retches—a sound often heard far across the ocean. That signals the colicky whale is preparing to reject what cannot be digested, and gift it to the sea. Not whale feces, not dragon spittle, but a lump of vomited-up intestinal refuse.

When first expelled, ambergris has a flabby consistency and a nauseating odor as it floats on the ocean surface. The lump often weighs three hundred to five hundred pounds, and lumps as heavy as a thousand pounds have been recorded. Newly ejected ambergris is a dark blackish color, but interaction with seawater and the atmosphere transforms it gradually to amber, silver-gray, golden yellow, and finally, in its sweetest stage, to a grayish white. At the same time, the stuff's clarity improves as its texture changes from waxy to pasty.

Over many years of floating on the sea, ambergris acquires a unique fragrance strangely tolerable to the human olfactory senses. Not only tolerable, the fragrance of aged ambergris has seduced millions of unsuspecting men and women whose partners have applied an ambergris-fixed perfume to strategic anatomical pulse points.

As a fixative, especially for flower-based essences, ambergris is said to have no match, its stability preserving fragrances for centuries without evaporation or change in translucence. Today, in Arab nations, ambergris is sold not only as a perfume fixative but also as an aphrodisiac and fertility booster. Some herbalists claim that inhaling the scent of ambergris offers an "estrogen cure," presumably intended for menopausal maladies.

In 1883, the New Zealand barque *Splendid of Dunedin* came across a 938-pound floating glob of ambergris, the single piece of flotsam valued at the time at about $250,000. In 1908, Norwegian whalers whose business was headed into bankruptcy discovered a glob of ambergris weighing 1,003 pounds, reportedly the largest single chunk of ambergris ever found. The whalers sold the ambergris for 23,000 British pounds and rescued their business. At the turn of the nineteenth century, the value of ambergris outdistanced that of gold; in the mid-twentieth century, it was at one point ounce-for-ounce worth eight times the price of gold.

Ambergris has been recovered from beaches and found floating in the ocean, but it is usually taken from the stomach of dead sperm whales. Once removed from the carcass and exposed to air, the stuff develops a waxlike quality, but it can also develop a pitchlike consistency or even become brittle. Scoundrels have tried passing off other substances as genuine ambergris, so buyers usually test the substance before swiping their credit cards.

The quickest test involves dissolving a small specimen in methyl alcohol (rubbing alcohol won't work) and then allowing it to cool. If the specimen crystallizes, it's ambergris. A more complicated test requires a needle or small piece of wire, and a flame. The needle is heated in the flame and then touched to the specimen, pressed into it about an eighth of an inch. A dark resinous liquid should form around the needle and the liquid should bubble as if boiling. Touching the liquid with a finger and pulling the finger away before the stuff cools should cause stringy bits to stick to the finger. When the needle, now coated with the melted substance, is heated a second time over a flame, the pitch should emit a smoky fume the same odor as the solid stuff, and then burn out with a clean flame. Finally, when the flame is extinguished, the smoke's odor should turn to that of burning rubber. This second test rarely fails to separate the genuine specimen from the fake. However, the most discerning buyers conduct additional testing, performed in chemical laboratories, looking for cholesterol and benzoic acids. If all goes well, the buyer may be confident it's the real thing.

John Singer Sargent's best-known Orientalist painting, *Fume'e d Ambre G*, 1880, depicts a woman, either ecstatic or astonished, I'm not certain which, standing over a pot of ambergris, inhaling its fumes. The painting was made during the height of the whaling industry, when every whaler hoped his knife would expose a pot of gold inside the big Moby Dick. Yet ambergris occurs in only about 1 percent of the sperm whale population.

Why, in this age of Swiss-made synthetic ambergris, would anyone covet whale yack?

Simply, because synthesized versions lack unique properties of the genuine article. When a synthetic version was introduced, the price of ambergris plunged, and yet certain perfumers, insisting that genuine ambergris imparts an unmatchable velvetiness to fine and expensive perfumes, still pay top dollar for the best-quality genuine article. Furthermore, perfumers say, its complex olfactory qualities—"amber, musk, animal, sea" and "a tobacco note"—cannot be reproduced. And ambergris as a fixative, preserving the fragrances of floral-based perfumes, cannot be matched. In fact, ambergris has preserved certain fragrances for centuries. Famous perfumes containing ambergris include Ambre Royale aux Fleurs No. 1114, Shocking by Schiaparelli, Arpege by Lanvin, Clandestine by Laroche, and numerous perfumes created by Chanel, Patou, and Guerlain.

In 2004, Bernard Pathé of the business Cadima Pathé, a purveyor of fragrances, claimed to have found a 130-kilogram lump of ambergris floating in the Pacific, near the Vanuatu atolls. At the time of his reported find, ambergris was selling for between $500 and $15,000 per kilogram, depending on quality. In late January 2006, news came out of Australia that beachcombers Leon and Loralee Wright had plucked a thirty-two-and-change-pound lump of ambergris off the shore. That lump has been valued at $295,000.

The auction Web site eBay lists about a dozen products claiming to be, or to contain, ambergris, including Bonne Bell Body Oil, which "has been the natural way for centuries in making perfume. It has a natural scent and its texture makes it the best in natural beauty enhancements." An ad from Morocco said: "This is genuine ambergris from Saudi Arabia and is fixes [sic] other scent." And from Bermuda, this: "Sperm whales frequent the waters around Bermuda and I have a lump of ambergris that is amber-yellowish in color and has a very distinctive pleasant smell. It is about 7.5 oz in weight. Bidding starting at $5."

The National Maritime Mammal Laboratory in Seattle maintains that ambergris is mainly harvested from whale carcasses, a viewpoint opposed by ambergris purveyors who sell to industry traditionalists. These ambergris merchants claim their "rare" ambergris is flotsam collected off a beach or found floating in the ocean, and they offer "certification" to that effect. The globs do occasionally wash up along seashores in China, India, Africa, New Zealand, the Bahamas, and Brazil.

While the 1985 Convention on International Trade in Endangered Species (CITES) protects the sperm whale, and indeed, many nations prohibit trade in ambergris, a controversy rages over technicalities in the CITES law as written. Opponents of the CITES law say it does not apply to the sperm whales' "urine, feces, and ambergris." Presently, U.S. commerce laws prohibit purchasing or selling ambergris. Possession of ambergris is prohibited by the Endangered Species Act of 1973, which includes the sperm whale. This means that U.S. manufacturers of fine perfumes today are not allowed to import or export ambergris, nor to sell perfumes containing it. Online purveyors of ambergris may claim the substance they are selling is "gathered, beach-washed product. No harm comes to any animal to obtain it," and "certified beach collected." The better side of caution recommends passing up the offer: A five-dollar purchase shipped from Bermuda to the United States could result in a ten-thousand-dollar fine and a holiday in Sing Sing.

But, wait. We're not through with Moby Dick.

As a child I suffered from anxiety attacks whenever my family went out for dinner. Inevitably I vomited at the table and was instantly removed from the party. Once at the tony Washington Athletic Club, an Easter dinner, I recall, when served ham with pineapple sauce, I immediately vomited across the white damask tablecloth. My Italian uncle was the only family member who spoke to me the remainder of the afternoon. The part that most mortified my mother was that the actress Vivien Leigh was dining at the next table and saw it all. And so I think I understand how a whale feels when the act of yacking draws attention. But the real case in point here concerns the candles used as decoration for such elegant dining, the Italian uncle who never heard of spermaceti, and how flotsam revolutionized the candle industry.

Candles have always represented class status, serving as a signal to visitors of their host's financial condition. Disgustingly pungent tallow candles in the eighteenth century decorated tables of lower-class homes and establishments. In literature as in real life, tallow candles reek of poverty and decrepitude.

In 1850, Charles Dickens's Mr. Booley (*Household Words*) recounts the Last Lord Mayor's Show in which a procession of tallow candlemakers was separated deliberately from the elephant, in order to spare the elephant the stink of the tallow makers: "After the Camel of Asia, came the Elephant of Africa. I found this idea, likewise, very pleasant. The exquisite scent possessed by the elephant rendered it out of the question that he could have been produced at an earlier stage of the Procession, or the Tallow-Chandlers, with their under Beadles, Beadles, and Band of Pensioners, might have roused him to a state of fury. Therefore, the Civic Dignitaries and Aldermen (whose noses are not keen) immediately followed that ill-savoured Company, and the Elephant was reserved until now."

Tallow, a hard form of sheep or cattle fat, was melted and molded into candles, which provided light and heat while the fat emitted its repugnant odor. Beeswax, the preferred material, made fine, sweet smelling candles, but like tallow candles, in the heat of summer, beeswax often turned soft and melted, making a mess of the dinner table.

Nevertheless, beeswax candles, from ancient Egypt to today, have enjoyed a fine reputation among the higher classes and the religious. Because beeswax candles burn cleanly and produce a pleasant odor, with little smoke and virtually no dribbling down a candle's length, they symbolize purity and are the preferred choice for religious rites of many faiths.

But it was a puzzling flotsam that struck like lightning over the poorly illuminated world. For eons a strange stuff had been spied drifting on the ocean surface and sometimes washed up along the shores. Buoyant, greasy, grayish in color, with a surprisingly pleasant odor, these mysterious globs—no, they weren't ambergris—fascinated sailors who discovered them bobbing in the sea, and while beachcombers generally found them too disgusting to handle, they nonetheless could not deny the blobs' attractive olfactory quality. People talked about it, much the way people today talk about mutton tripe—in hushed, cautious tones.

History records that eighteenth-century whalers finally identified the floatable globs as the same substance that they found in a sperm whale's unique skull cavity, which they called spermaceti. Then someone—I'm betting on a whaler's wife—discovered the blobs might actually be good for something. Once the blobs had crystallized, they formed a fragrant wax, not quite suited to the bikini line but fragrant and long-burning, and therefore perfectly suited to candle making.

The sperm whale derives its common name from its spermaceti organ, which comprises the greater mass of matter inside its skull. The organ is a waxy glob that possibly provides some buoyancy control, assisting the whale in diving and ascending. With a change in blood flow, the whale can alter the substance from a solid wax to a liquid state. In this theory, when solid, the organ becomes denser, allowing the whale to sink; when liquid, it is less dense, providing more buoyancy. A second theory suggests the spermaceti organ is used for echolocation, allowing the whale to focus and control sound waves. Both theories may prove correct.

Literature introduced the world to spermaceti in Hermann Melville's *Moby Dick*: "Moreover, as that of Heidelburgh was always replenished with the most excellent of the wines of the Rhenish valleys, so the tun of the whale contains by far the most precious of all his oily vintages; namely, the highly-prized spermaceti, in its absolutely *pure, limpid, and odoriferous state* [my italics]. Nor is this precious substance found unalloyed in any other part of the creature. Though in life it remains perfectly fluid, yet, upon exposure to the air, after death, it soon begins to concrete; sending forth beautiful crystalline shoots, as when the first thin delicate ice is just forming in water. A large whale's case generally yields about five hundred gallons of sperm, though from unavoidable circumstances, considerable of it is spilled, leaks, and dribbles away, or is otherwise irrevocably lost in the ticklish business of securing what you can."

And so flotsam blobs transformed candle making. The resulting spermaceti oil, when burned, elicited that pleasant odor and, too, spermaceti wax was harder than tallow or beeswax, producing longer-burning candles. By the late eighteenth century, spermaceti wax had set the standard for candle making. Beeswax remained the candle of choice for religious ceremonies. Tallow was reserved for the lower classes.

Although spermaceti trade is, like ambergris trade, prohibited by CITES, the market for spermaceti is worldwide and healthy. The stuff is as easily purchased as ambergris on eBay, but spermaceti today usually is physically extracted from captured whales, and no captured whale survives the extraction. Anyone offering spermaceti oil for sale likely obtained it from an illegally captured sperm whale. Rarely, anymore, does spermaceti wash up.

One final note on the subject of spermaceti: When I asked my Italian uncle if he had ever heard of spermaceti, he replied, "That's a Sicilian dish. We northerners aren't so crude."

Mrs. Stramanos's Wonderful Flotsam Jewelry

I know a woman who covers herself in flotsam jewelry. Her earrings, necklaces, bracelets, and rings are all made from ancient flotsam washed up on beaches in and around Riga, Latvia, where her husband is from. He's a successful neurologist with what Americans call *disposable income*. Apparently all of Dr. Stramanos's disposable income is invested in his wife's jewelry. She is a tall, broad-beamed, refined Amazon who lopes when she walks, probably a result of the combined weight of her jewelry. Each time I see her, Mrs. Stramanos has added more flotsam around her neck, her wrists, or on her ears. She is beginning to resemble a glorious noble fir decorated for Christmas. The most amazing thing about Mrs. Stramanos is that she can tell you where each piece of her flotsam jewelry originated before it went afloat on the Baltic Sea eventually to wash up in Latvia.

Mrs. Stramanos wears amber. Some of her necklaces include amber beads the size of golf balls. Translucent and millions of years old, the cognac and honey-colored amber chunks, carefully cut and polished, reveal tiny bits of prehistoric plants and insects, even little hairlike threads inside. One pendant hanging very near her jugular vein contains an ancient mosquito with blood in its stomach. If the mosquito ever escapes, God help Mrs. Stramanos. It looks like it wants to bite Mrs. Stramanos, but she calmly reassures me that creatures trapped in this ancient flotsam will never again fly or crawl or bite. Mrs. Stramanos is a walking textbook on the subject of amber flotsam and its trapped prehistoric flora and fauna.

The poet Alexander Pope in "An Epistle to Dr. Arbuthnot" (1735), wrote this:

Pretty! In amber to observe the forms
Of hairs, or straws, or dirt, or grubs, or worms;
The things, we know, are neither rich nor rare,
But wonder how the devil they got there.

Fossil amber, Mrs. Stramanos tells me, when it's rubbed against a cloth, becomes charged with negative electricity. Baltic peoples believe that, like ocean-borne negative ions, this negative electricity is a good thing. Not only Baltic folk: New Agers, shamans, Wiccans, and soothsayers also ascribe supernatural properties to amber. A recent eBay auctioneer, hawking "Golden Amber Rough Organic Gemstone" from the Baltic Sea under the moniker "Future of Light—Feel Yourself Here and Now," claims that amber "dissolves negativity and depression," is a "Disintoxicant and painkiller. Helps us to find greater peace and harmony, transforms negative energies in positive [sic] and develops altruism. It is a symbol of personal power, health, virtue, happiness and joy of living. Purifies the living and working spaces and protects from radiations." And here is the one and only fact upon which eBay amber merchants and Mrs. Stramanos agree: "In contact with the skin its electromagnetic properties create a field that tends to heal the area around it."

Snicker if you will, but whenever I approach Mrs. Stramanos at a cocktail party or other social event, I immediately sense an electromagnetic irradiation of all negative aspects of my being, notice increased purification and a harmonious attitude, and if I have a stomachache or similar internal or external malady, it immediately disappears.

Fossil amber is the fossilized resin of *plinus succinifera*, coniferous trees, from the Eocene period about 40 million to 55 million years ago. Mainly found washed up on beaches of the Baltic Sea or floating upon its surface, amber is also mined in open pits, but it's the flotsam amber that's most prized, I suppose because a gift from the sea carries more romance than pit-mined treasure. For thousands of years, Mrs. Stramanos explained to me, collectors have prized chunks of amber whose honey, golden, or cognac translucence not only emits a warm incandescence but also reveals whole insects, or flower blossoms, or swaths of pollen of prehistoric vintage trapped inside and perfectly preserved. Mrs. Stramanos has an amber bead enfolding a rare intact prehistoric flower blossom, a precursor, I think, to edelweiss. It's quite a treasure. But it's rare insects that really excite the cognoscenti. A fossilized rare species of gnat, for example, can send thrills up an entomologist's spine and trigger instant inflation in the amber market. Mrs. Stramanos has an amber cabochon ring whose stone encases a rare beetle species.

I ask Mrs. Stramanos how the insect got inside her cabochon. "Millions of years ago, when resin dripped down a tree's bark, insects would get trapped," she tells me. "When the resin set, the insects became encased for eternity. This has been a boon to entomology. Of course, the same is true for flowers and other plant materials, contributing to the knowledge of botany and other natural sciences."

One of her favorite subjects is the Victorian fascination with amber. "The Victorians revered amber, not only for its beauty, but for its healing powers," Mrs. Stramanos informs me. "Especially the Brits. The English bourgeoisie made a sport of acquiring really big chunks of amber with all sorts of interesting stuff trapped inside. They used them as paperweights, or just as conversation pieces, and they set them on their parlor tables to impress their guests."

Mrs. Stramanos has made the pursuit and knowledge of amber her life's work. This may explain why Dr. Stramanos always seems to wear a wistful, slightly wan expression. The poor man has lost his true love to his native land's most revered flotsam. He is literally hidden behind her now when they appear at social gatherings, with Mrs. Stramanos bedecked in her wonderful flotsam jewelry.

Amber's components include approximately 80 percent carbon, 10 percent hydrogen, and small quantities of sulfur. Its hardness registers 2 to 2.5 on Mohs scale. Think of a substance not quite as hard as a pearl. Fossil amber varies in color depending on what materials it contains. Black amber contains tree bark or bits of fossilized plant material from forest floors. Brown and green amber contain moss. The clearer shades of cognac, golden yellow, and even pale ivory contain mostly resin and because of their lightness and translucence, show off the captured detritus. Rare specimens of blue and cherry amber have been plucked off Baltic beaches, and these are highly prized.

I ask Mrs. Stramanos if amber can be found anywhere besides the Baltic region. She sniffs. "People will say so," she demurs, "but it's not really amber. The stuff that comes from the Dominican Republic, or from New Jersey? That's all much younger than Baltic amber, much less interesting to the knowledgeable collector."

"New Jersey?"

Mrs. Stramanos rolls her eyes. "So they claim."

When found on the Baltic shore, amber can seem like a lump of plastic with a rough, weather-beaten character, not the purveyor of supernatural golden light that will reveal itself once it is properly cleaned and polished. A good test for amber's authenticity is to put a lighted match to it. While plastic gives off a putrid chemical odor, amber releases an aromatic resinous perfume. My Scots grandmother used to wear a perfume called Tabu, which gave off distinctly amber bottom notes. I tell Mrs. Stramanos this.

"Of course," she starts, "amber essence is somewhat of an aphrodisiac."

I stop her right there. I can't bear even an insinuation that my grandmother dabbed aphrodisiacs on her décolletage.

Both ambergris (of whale yack origin) and amber are among the oldest known substances traded between Asian and European merchants. Both substances have for millennia been burned as incense, and applied to the body or swallowed for medicinal reasons. Sailors have made fortunes collecting the sea's most exotic flotsam.

The demand for amber jewelry is so great that a process called pressed amber was developed to spread the material around. In this process, small bits of true amber otherwise unsuitable as jewelry are heated to nearly 600 degrees Fahrenheit (around 320 degrees Celsius) and pressed together into plate or cylindrical shapes, from which objects are stamped or carved out. Sometimes the jewelry maker will sprinkle gold or silver flecks into the liquid amber before it is pressed. These flecks impart the false impression of captured organic material. Much of the amber jewelry for sale today is manufactured from this pressed amber technique. The Chinese melt honey-colored resin and cleverly insert modern-era bugs whose corpses they have tortured to give a prehistoric appearance. They sell these miniature artifacts on eBay for ninety-nine cents a pop. They are tiny frauds, worth the price. The Chinese, Mrs. Stramanos tells me, sell so many of these fakes that if the Cultural Revolution's Away With All Pests campaign wasn't successful the first time around, surely by now all of Guangzhou's flies have been captured in amber pendants for sale on eBay.

Chunks of real amber containing insects or flora proliferate on eBay auctions, but these are mostly processed amber or fragments containing boring species of flora and fauna; few measure up to Latvian standards—the highest indicator of quality. None of the amber decorating Mrs. Stramanos would ever be traded online.

I have walked Latvian beaches hoping to find even the tiniest chunk of amber. I don't care if it holds an insect leg or rare flower petal, or botanically interesting pollen; a tiny speck of prehistoric pollen would suffice. But as Mrs. Stramanos explains to me, amber flotsam has become harder and harder to find. "Most of the really rare stuff," she says, "has already been snatched up. Sadly, the Baltic has given up most of its amber."

(But why should she worry? She's wearing half the world's amber supply around her neck.)

"But every once in a while," she adds as if to comfort me, "a beachcomber can find a piece of true amber on the Baltic beaches."

Yeah, sure, like one day the tide will deliver up my own special chunk of amber here on the shore of this Latvian fishing village, and it will land right at my feet and be mine to treasure or trade.

"Maybe you should try Estonia," suggests Mrs. Stramanos.

"Already tried," I say. "Any other ideas?"

Mrs. Stramanos is a patient woman, a good thing considering our size difference. Her broad hand drops onto my shoulder, its pinky finger sporting a new acquisition, a honey-colored cabochon loaded with exotic prehistoric detritus. "Listen, dear," she says, the way a pro golfer explains double bogies to a novice, "you need to know what you are looking for. And then you need to know which beaches to search, and how the currents flow, and how to watch the tides, and where to search among the wrack."

I scratch the side of my face. "Let me ask you this," I say. "Where did you find all your amber?"

Mrs. Stramanos's fingers caress her amber-clotted breast. "These? Why, I found them in Riga," she says, "in a little shop in Priditis Street. They have a marvelous catalog order service, too."

Shell Lust

Manhattanite Sarah Soffer collects old bones. Sarah is especially fond of piscine and marine mammal remains washed ashore on coastal beaches. She's a photographer and enjoys arranging the remains against simple backdrops and taking their pictures, sort of intimate postmortem portraits. I once tried interesting Sarah in seashells but she turned her nose up. "It's not the same as bones," she sniffed. "Bones are more personal." I wonder about that. Seashells are bones worn on the outside.

If humans wore their skeletons on the outside, would they be more enamored of themselves? If, like most mollusks, we went around in bony armor to protect our soft flesh, our tender organs, would we flaunt our bones? Wear them haughtily and be less fearful of predators? What would we look like as exoskeleton creatures?

Clams, scallops, snails, and other marine mollusks have an amazing ability to disguise and protect themselves with

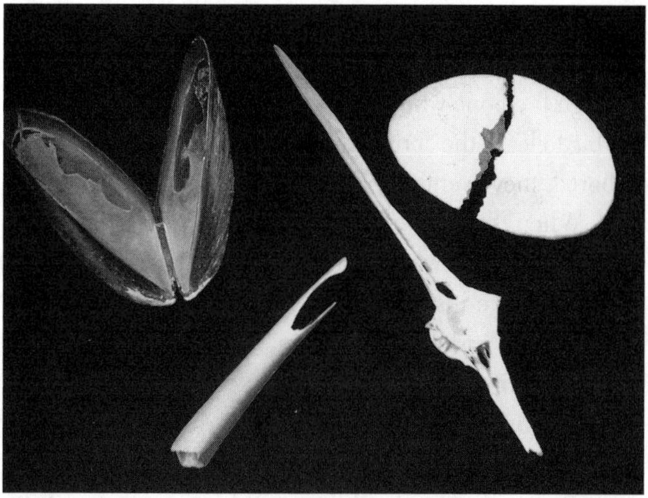

Eco and ecto skeletons.

their skeletons, some so intricately constructed, so architecturally ingenious that great architectural projects have been modeled after them. Take the *Euplectella aspergillum*, for example, better known as the Venus flower basket. If a VFB washed up at Sarah's pretty feet, I guarantee she wouldn't spurn it.

The Venus flower basket is a sponge with a hollow cylindrical glass skeleton, generally ten to eleven inches long, a fragile phallic scaffolding constructed of intricately latticed woven glass. It attaches itself to the ocean floor with glass fibers that resemble the finest spun angel hair. Its shaft resembles tulle netting surrounded by swirled glass bracings. It is one of the world's rarest and most prized bits of flotsam, found only occasionally washed up intact on the beaches of Cebu island, in the Philippines, where it lives at depths of fifteen hundred to fifteen thousand feet in the Western Pacific Ocean. Frequently, a pair of tiny shrimp will enter through the bone-white latticework and grow until they are too big to leave the "bridal chamber"; thus encased, certainly bored, they eventually mate. What else is there to do in jail? When their offspring hatch, the little brine swim through the latticework openings, but the larger parents are trapped for life inside the structure. In parts of Japan, the Venus flower basket is a traditional wedding gift, meant to represent a lifetime commitment. Interpret this as you will.

Western architects have begun studying the design of the Venus flower basket, hoping to translate its inherent strength into the structural design of high-rise towers. Joanna Aizenberg, a researcher at Bell Laboratories/Lucent Technologies in Murray Hill, New Jersey, published a paper in the journal *Science*, saying the VFB's elaborate structure represents "major fundamental construction strategies such as laminated structures, fiber-reinforced composites, bundled beams and diagonally reinforced square-grid cells." In other words, high-rise towers like I. M. Pei's Hong Kong Bank Building with its diagonally reinforced beams, London's Swiss Tower, and Paris's Eiffel Tower all imitate nature, except, unlike the Venus flower basket, they provide exits. Scientists have also discovered that the VFB and other bottom-dwelling sponges in the genus *Euplectella* possess filaments able to conduct light "as well as or better than the best fiber-optic cable."

I own two Venus flower baskets. I keep them wrapped in tissue paper inside cylindrical boxes inside an Asian chest. Every once in a while I carefully remove these fragile

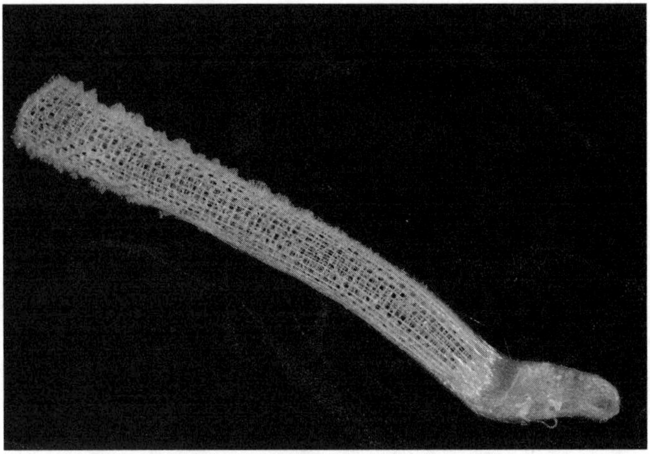

The Venus flower basket was once a living organism.

skeletons, marvel at their strange beauty, and thank God I'm not a trapped shrimp.

My friend Sarah would certainly appreciate the bone-like appearance of the Venus flower basket. As for the sponge's ability to trap mating pairs, Sarah would sniff and say, "It's a matter of interpretation." Which brings us back to the seashell, these beached skeletons of mollusks, sea animals who built their bones for armor.

Think of it: a skeleton worn on the outside. Once the organs fail, once the flesh succumbs, all that's left is bones, and, frankly, humans have nothing to brag about, being somewhat the cuttlefish of mammals. Compare a typical everyday human skeleton with the shell of a Tahitian heart cockle, or a suggestive violet-lipped *Cassidula nucleus* from the Philippines, or yellow-lined *Strombus fasciatus* from the Gulf of Aqaba, in Sinai, Egypt, or the African turbo with its mahogany veneer over mother-of-pearl, or even the simple abalone, the plate limpet, for criminy sake, or the fighting conch. How dare we humans brag about our bones?

Because most mollusks wear their skeletons on the outside, providing shelter and defense, beachcombers often forget that something might be living inside these shells, which have always enjoyed currency in human cultures. Traded as money, used for medicinal purposes, talismans, as ornaments or *objets d'art,* seashells have sparked human curiosity ever since Cro-Magnon man stood on the strandline and marveled—if Cro-Magnon was capable of marveling—at them rolling in on the tides.

The tropical marine mollusk *Cypraeidae*, or cowrie, develops a brightly colored shell with a long central column that works like a human spine. Cro-Magnon man wore the cowrie shell to stave off sterility. Apparently it didn't work so well. The pit dwellers of prehistoric England used cowries for lucre, as jewelry, and as talismans, as did the Saxons in Germany. In pre-dynastic Egypt, cowries were named for the Latin *Cypraea*, which has the same root as Cyprus, the birthplace of Venus, or Aphrodite. One look at the entrance to the cowrie shell's erotically suggestive interior reveals why. Marco Polo, on the other hand, visualized cowries as "little pigs"; thus he dubbed them *porcellana*, a word eventually used by Europeans to describe Chinese porcelain, the delicate surfaces being reminiscent of the smooth cowrie shell.

In ancient China, cowrie shells were incorporated into burial rites: While a commoner's corpse had its mouth filled with rice, an emperor was buried with cowries stuffed in his mouth—nine cowries being the maximum, and I suppose it depended on how big the emperor's mouth was. Feudal lords merited seven cowries in the mouth, while high officers only got to suck five, and ordinary officers tucked three. An exception was made for commoners of great wealth; a latter-day Donald Trump or Bill Gates, for example, received a small cowrie on the last molar on each side of the mouth to ensure the departed soul had enough to eat and spend in the next life, implying, perhaps, that certain privileged cowrie suckers *can* take it with them.

In Bengal, cowries were small change, not a king's ransom. In East Africa, Arabs used cowries both as money and female fertility charms; they were also thought to deflect the evil eye. Burial sites near the Caspian Sea in the Caucasus Mountains reveal the use of various families of mollusk shells as lucre to help transport souls. In Africa, especially in Uganda in the nineteenth century, cowries were traded as currency.

The Old Testament describes Adam and Eve fleeing Paradise, clasping fig leaves to their private parts. I'm not sure why their beaded jewelry wasn't mentioned. It's a fact that the beads came first. Even before the first recorded sinners clasped ficus to genitalia, Cro-Magnon was decorating his body with beads made from seashells washed up near his seaside cave. Cro-Magnon males made beaded jewelry from shells of univalves or gastropods—snails, whelks, and conchs-strung on plant vines. The craft of bead making developed parallel to tool fabrication, and eventually these prehistoric hippies figured out how to slice abalones into small pieces that revealed the excellent character of the shell's mother-of-pearl lining. Cowries, too, were sliced for their inner beauty. Much later, professional bead makers would agree that the naturally worn surface of an Indonesian blue-backed cowrie shell cannot be matched by artificially induced versions.

Ever wonder what the world's oldest professionals were paid for their services? Or why Cro-Magnon man traipsed around wearing stacks of beaded necklaces? A simple leap of logic explains why he frequented the strand, beachcombing for cowries.

Even today, *Homo sapiens* derive pleasure from stringing flotsam around their necks and limbs, hanging it like tree ornaments from their ears. Mrs. Stramanos is just one example of this phenomenon; think of Don Ho and his Hawaiian shell chokers.

Linnaeus, when developing his mollusk classification system, named the quahog Mercenarius (Venus); and America's first form of exchange for value, *wampum*, from the Algonquin word *wampumpeag*, meaning a string of white beads, was made from the quahog shell. While less prized wampum was carved from white shells of whelks and conchs, usually from a shell's central column, the most valued wampum was fashioned from quahog shells, an Atlantic Ocean native that bears a patch of deep violet on the interior face. "Council wampum" was carved by Native Americans, such as the Wolf Clan of the Mohawk Nation at Akwesasne around Lake Oneida, New York. These were later imitated by European factories, but the production-line versions were not as prized. Among Native Americans, wampum was highly symbolic, offered in kinship affirmation rituals and condolences. French explorer Jacques Cartier, in 1535, observed an analogy between the uses of wampum by Native Americans and the Europeans' use of precious metals like gold and silver. Thus, at least to the European mind, wampum developed a reputation as "currency."

Today, the quahog is less prolific, having been overexploited. Pollution in the near-shore waters where quahog mollusks once thrived has caused the distinctive violet patch to shrink. Still, if you've never seen the violet insides of a quahog shell, well, you simply haven't lived.

Mollusks have been collected off beaches for their colors alone. For more than thirty-five hundred years, humans have squeezed dye from mollusks of the genus *Murex*, using it to color their bodies and their clothing. Early records date back to Crete, but some historians believe Neolithic man was already using mollusk juice to decorate himself. Because dye making was labor intensive and too costly for the commoner—three hundred pounds of liquid Murex dye for every fifty pounds of wool—the color it produced became associated with royalty. Antony and Cleopatra's sails were tyrian purple, with dye derived from the same mollusk family mixed with the red dye produced from whelks. During his reign, Nero was the only Roman permitted to wear tyrian purple. Incredibly, mummy wrappings dyed with mollusk purple have retained their color over several thousand years.

Central Americans in the sixteenth century learned to squeeze purple juices from mashed *Purpura patula* snails they collected on beaches. During the Spanish occupation, the dye derivative was shipped to Spain to dye noblemens' cloaks and vests. Back in the colonies, the shell mashers eventually figured out that smashing snails endangered the animal population, so they taught themselves to blow into the snail's shell, causing the dye to dribble out. The snails were replaced in their habitats, free to reproduce, arguably one of humankind's first efforts at conserving an endangered species.

Even more fabulous than the whelk and mollusk dye is what ancient Mediterranean divers discovered on the seafloor: the noble pen shell, perhaps the most fascinating article ever brought up from the sea. The noble pen mollusk anchors itself to the seafloor with fine golden threads about two feet long, strong enough to hold the animal steady in the wake of powerful currents and undertow. Jason seeking the Golden Fleece may have been pursuing the noble pen's threads. From ancient times, these exquisite golden threads have been sewn into the clothing of the wealthy classes.

It's a fact that fashion designers for millennia have aspired to the shapes, textures, lines, and weave of seashells. Like architects studying the structural design of the Venus flower basket, certain organically oriented clothing designers strive to decorate the human body in Neptune's colors and patterns. Scientists envy sponges and mollusks, more so because they can't yet exactly match their designs, but with the advances in 3-D software technology, we may soon crack the code. It is entirely possible in coming epochs that we humans may see the advantages in turning ourselves inside out in order to survive. More fantastic evolutionary moments have occurred.

A seashell is most vibrant when the animal is still living inside it. Once it dies, the shell gradually begins to deteriorate and fade. Exposed to sunlight, seashells lose their beauty and luster, and caring for them over long periods of time requires a conchologist's expertise. Seashell guidebooks often warn beachcombers to check federal, state, and local laws protecting marine life before going shelling on a beach. They also, with a straight face, warn against harvesting shells whose animals are still alive and residing within. I have often wondered how this applies to the turbans and the conchs and other architecturally complex shells. Does one knock? Ring a bell? Shout, "Anybody home?"

I remember as a child that a female adult in my family, perhaps an aunt, frequently sported a particular necklace—a spiky fire-orange choker that looked like, if she cocked her head, it would puncture her jugular vein. No one knows how many necklaces, bracelets, hair ornaments, earrings, rings, Bedouin prayer beads, carvings, calcium supplements, and kitschy souvenirs have been created since humans first began mining coral reefs, but over the millennia of human accessorizing, with perhaps the single exception of the Cro-Magnon fetish for shells, coral has enjoyed top billing among fashionistas, even being hailed recently as a summer must-have accessory.

Coral washes up on the strangest beaches. I plucked my first piece of coral off the tide at Myrtle Edwards Park in downtown Seattle. Little white spiky thing, not nearly as lovely as auntie's jugular wrap, still, coral. Since that first find, I've collected small bits of white coral from beaches in California and Oregon. And this is how eBay's coral hawkers claim they acquired the precious fire-orange and deep red stuff they offer online. Not so. Yet those items of jewelry and souvenir kitsch represent only a tiny fraction of coral reef

Some beach shells seem to stare at you.

demolition. A flotsamist, therefore, who happens upon pieces of coral on a beach should thank the stars for such good luck and treat it kindly, for once it was alive and thriving and serving the kingdom of the sea.

Barnacles

The barnacle is a curious bit of organic flotsam, and honestly, one of my favorites. Charles Darwin spent eight years studying barnacles, and his monograph on them is still required reading for marine biologists. The two-volume masterpiece, published in 1851 and 1854, describes the barnacle's curious physical features, its free-swimming young, and its feeding habits. Whether or not they've read Darwin's masterpiece, accomplished beachcombers know barnacles intimately, either from nasty cuts to bare feet or from finding the perfect glass ball fishing float washed up and encrusted in hitchhiking barnacles. For centuries barnacles were thought to be mollusks related to oysters, clams, and mussels because they seemed to possess a shell. Darwin and his contemporaries, however, had observed the barnacles' wandering offspring, placing the animal in the order of crustaceans, like shrimp and lobster. What had

The horseshoe life preserver washed up long ago is now covered with barnacles.

been misnomered "shell" was actually layers of calcareous plates the animal secretes as protection.

A barnacle's life is not necessarily boring. True, if you were stuck to a boulder that never moved from its position on the tide line, life might prove a yawn. But imagine being attached to a cargo vessel, traveling across the oceans to exotic ports where you mix and mingle with other barnacles stuck to other cargo vessels. Or how about riding a whale? This is as adventurous as barnacle life gets, but I know humans with duller lives. At least humans can perambulate; barnacles, once attached, are stuck for life. Fused by its own strong glue, to driftwood, or rocks, a crab or whale's body, or to a ship's hull, the barnacle goes where its ride goes. Necky goose barnacles are fond of sticking themselves to wood pilings or ships' hulls from which they dangle headfirst into the water. Their bodies are confined between plates shaped more or less like a goose bill, and legend had it that they would hatch into "barnacle geese," thus the goose moniker.

The more common acorn barnacle attaches its neckless plated armor directly onto whatever it chooses to ride; cemented headfirst, its feathery feet protrude, waggling for food. Spawn is ejected from the hermaphrodite's elongated penile tube in hopes of fertilizing its neighbor. Such is barnacle sex.

Barnacles damage hull paint and often clog drainpipes opening into the ocean. They have justly earned the sobriquet "most prolific fouling marine animal." And all along we thought that title belonged to humans.

Never underestimate the sticking power of a barnacle; cements like Superglue were first developed from studying barnacle glue. Patent or not, I would personally be loathe to have my glue taken, and, anyway, barnacles deserve far more respect than they currently enjoy. Research has shown that barnacles possess a remarkable ability to accumulate concentrations of poisonous metals, suggesting potential applications for cleaning up pollution. For decades marine biologists have used barnacles as a measure of pollution levels in saltwater. Flotsamists know that barnacles stuck to flotsam indicate the object has spent a long time at sea. This is an important clue when attempting to identify how far and how long a piece of flotsam has traveled on the ocean. And, of course, like every sea animal, the lowly barnacle has found a place in haute cuisine. Next time you see barnacles on the menu, think glue.

Ancient Flotsam Cures and Cosmeceuticals

Ancient peoples believed that gods sent healing remedies from the sea to those who respected it. A peek into an ancient pharmacopoeia offers examples of medical treatments involving flotsam.

Seahorses, which travel just beneath the water's surface, have always been collected, both for ornamental and medicinal use and by children for purposes of torture. Ancient Greeks employed seahorses in an attempt to cure cancer, and Europeans believed seahorses cured urinary incontinence and improved the flow of breast milk. Children still collect seahorses to add to their "sea rodeo" collections, and this is where the torture enters. I once witnessed a five-year-old boy playing sea rodeo with freshly harvested seahorses from the Indo-Pacific. The boy had

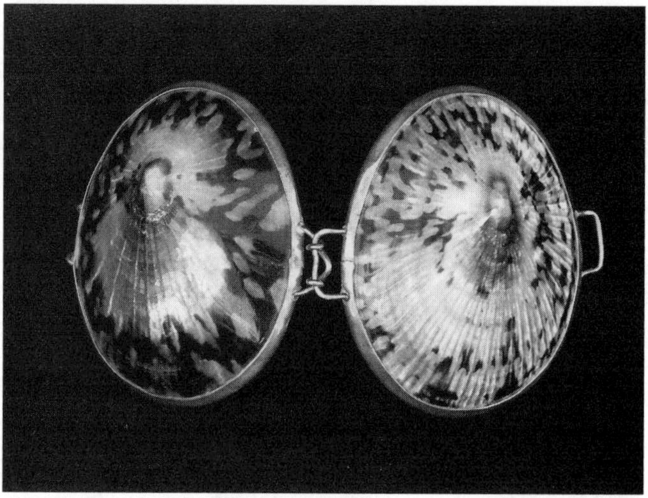

Limpet shell and gold snuff case from Dover, England.

fashioned a miniature lasso from dental floss. I needn't elaborate. Pity he didn't realize that an intact seahorse washed up on the beach is thought to bring great fortune to its finder.

Pearls are also believed to lure good fortune. Tahitian black pearls are particularly effective in curing female depression, and just now I'm really, really depressed. Pearls also were used to reduce fevers and to treat various organ maladies. Crushed pearl mixed with crushed coral was thought to cure tuberculosis. Today, crushed pearls are used in Chinese and Japanese cosmetics made for improving skin texture and clarity, or at least the manufacturers' bottom line. Asian women swear by crushed pearls, though, like this writer, they're fonder of the uncrushed variety. Persians believed pearls cured insanity, and I'll second that.

On a recent layover in Hong Kong, I devoted an afternoon to pearl creams. My old pal Simon Chu, who manages a pearl export establishment, pointed me to the best apothecaries, and when all was said and done, I came home with a product called Make Best Face for Husband, a rich, thick, pearl-infused emollient produced by a factory in Macau. The instructions advised using only on face and hands, as "ingradients may allergic other body parts." In the end, Make Best Face for Husband gave me a rash around the hairline and nostrils, but I swear I saw crushed pearls in the formula.

Abalone shell, the English believed, made short shrift of corns. Here's the recipe: "Squeeze lemon juice over mother-of-pearl buttons 2x a day for one week, until the buttons turn to paste. Spread on corn, cover with bandaging. Repeat once a day until the corn detaches from the foot." Next time you lose a button try checking your neighbor's foot.

Seaweed, marine algae, stranded fish, and invertebrates are listed in ancient pharmacopoeias. Egyptian, Chinese, Greek, and Roman physicians combined ministration of organic matter washed up on their beaches with prayer to heal the sick. Brilliant. They actually slopped on the raw, slimy stuff straight off the wrack, and here we discover the origins of La Costa and other spendy spas. Today, native healers and shamans compete elbow-to-elbow with spa owners along the wrack line, collecting medicinal and cosmetic miracles piled up on beaches—jumbles of green and brown seaweed, and kelp so tangled it snags seahorses, mussels, crabs, and other marine life, whose leaves or roots or skins found their way into the apothecaries of every ancient seashore community in the days when every part of a plant or animal was sacred. But I'll bet if they had Aleve and Botox back then, the wrack would have languished where the tide left it.

In Europe during Victorian times, seawater treatments became all the rage as thalassotherapy, and spa aestheticians applied seaweed poultices directly to their clients' skin, then bathed the skin in seawater to help detoxify and nourish the body. Because of its amazing softening properties, seaweed has for centuries been used in traditional Asian pharmacology, applied to tumors and nodules. As an anti-inflammatory, seaweed has been applied to swollen lymph and thyroid glands, and when taken orally seaweed is believed to reduce swelling of lymph nodes and, more recently, to generally help combat symptoms associated with HIV and other autoimmune disorders and certain chronic allergies.

Chondrus chrispus, Irish moss seaweed, traditionally was used to remedy digestive problems like diarrhea, and to treat dysentery, gastric ulcers, and the common cold. Both green and brown seaweed—ubiquitous flotsam debris—contain minerals and other nutritive properties and have for centuries been a dietary staple, as common as lettuce and other land-grown vegetables, and, if possible, less palatable. "Eat your seaweed, young man, or you'll never make samurai."

Irish moss is a gloriously versatile flotsam that is also used as bedding. I've lain on a moss bed in a hotel in Rome, which shall go unidentified only because it is one of those preciously held secrets one can't afford to advertise for fear of creating a tourist wave that would spoil its cachet. I can state without hesitation that this hotel's Irish moss bed is even more comfortable than Leonard Bernstein's bed.

They say never kiss and tell. I didn't kiss, but I'll skip the details other than to report that I spent a night in Mr. Bernstein's bed, and in that downy nest enjoyed the second-best night's sleep of my life. In fairness I must divulge that the maestro and I did not share his bed—I had it all to myself. On a scale of 1 to 10½, the maestro's bed ranks 10, while the Irish moss bed in Rome ranks 10½. But getting back to flotsam, picture this:

A sunny springtime in Connecticut. Midafternoon. The swimming pool at Leonard Bernstein's country residence. I am lounging poolside, savoring robin song, when I notice a stick adrift in the swimming pool. A whisper of breeze had stirred up a current on the water's surface, causing the stick to flow in and out of lacy shadows cast by a shade tree. Is that a stick, I asked myself, or . . . good heavens, could that be a baton? Was it possible that somehow the maestro's baton had fallen into the swimming pool, to travel like so much flotsam from the diving board to the stepladder, trapped in heartless currents, its delicate balance thrown off by the aquatic insult? I removed my sunglasses and stood up. Straightening my bikini bottoms, I marched barefoot toward the pool. The melody to "The Rumble" entered my head and I must have been humming out loud because the startled robin stopped singing and flickered off. I dove into the pool and swam for the sacred flotsam. As my fingers grasped the slender stick I at once perceived the flotsam was more suited to the robin than to the maestro. At that moment my host ventured out from the house. "I didn't know you were quite so compulsively a neat-freak," he said. "How about another martini?"

Bladderwrack, *fucus vesiculosus*, a floating seaweed, grew popular in nineteenth-century North Atlantic coastal regions as a treatment for goiter and obesity. Now my interest piques: Did they really distinguish between goiters and obesity, or did they think dad's pork belly was a goiter? Anyway, the claim itself is suspect—who knew from goiters back then? I think the bladderwrack claim is a bogus attempt to promote sales of *fucus vesiculosus* by Shopping Channel shamans who adopt transparent Scottish brogues and claim the cure was passed down in family lore. Besides, no wrack I've ever seen has a bladder, and I've even looked up a kelp's hose.

Japanese have traditionally embraced—or rather, imbibed—sea vegetable diets for their laxative qualities. Arame, hijiki, kombu (a kelp), and wakame have all been shown to eliminate toxins from the intestines because they contain an alginic acid, which binds to heavy metals in the intestines, making them indigestible; the toxins then glide from the body along with the alginic acid. And today, even Morpheus wears a seaweed cure: a new brand of seaweed fiber pajamas that detoxify while you sleep.

Poseidon's Drifting Limbs

Twelve thousand years ago, Iceland's first settlers chopped down the island's trees for building material and fuel. So thoroughly did they denude the land of trees that they came to depend on Siberian driftwood for home construction and fuel. I've visited Siberia, have personally eyeballed Siberian tree trunks, in fact, stumbled into one, after too much Inupiat white lightning. Siberian trees are not, well, girthy. They are thin as a Finnish rock star, and I venture to say, were probably no thicker twelve thousand years ago. Siberian driftwood can't be all that desirable as log cabin material.

Eventually someone in Iceland's denuded taiga got the eureka of growing, hopefully, fatter trees from seedlings arriving via Siberian driftwood, but in the meantime, sheepherding was introduced, and the dimwitted ruminants munched out the tree saplings before they could mature. Decades later, still dependent on Siberian flotsam, someone fenced the sheep in, and the tree saplings eventually matured into girthy construction-grade timber. Iceland's strandliners still covet wood flotsam, though, just in case the sheep get loose.

An amazing quality of driftwood is how far it can travel without deteriorating. Branches of Pacific Northwest Douglas firs have been found on Polynesian beaches. Rare Chinese woods have been collected along the Oregon coast—not only remnants of shipwrecks, but also parts of uprooted trees gone adrift. Hardwoods from the tropics have washed up along Maine's coast, but to the amateur beachcomber's eye, the wood is indistinguishable from other driftwood lining the beach. Experts know the difference, and they have made startling discoveries that have led to an understanding of the ocean's currents and weather patterns.

I met Lonnie Rose in Garibaldi, Oregon. Lonnie is a woodworker. I am not sure if that means a carpenter or a sculptor, and I never had an opportunity to find out because when Lonnie starts talking about driftwood flotsam, he's a runaway train. He's a whole-body talker, moving in rhythm to his story, his hands waving, his hips swaying—you'd think he was dancing—his mesmerizing tales of driftwood flotsam painting a picture of every piece of wood he's ever claimed off Oregon's wild beaches. Lonnie Rose has literally fenced himself in with flotsam. A fence he built of driftwood flotsam surrounds his blink-and-miss home on the precipice of U.S. Highway 101 running through Rockaway Beach. But Lonnie's favorite flotsam is kept indoors because it is too incredible to appreciate in peripheral vision.

It's a horse's head formed in the grain of a cedar log. Lonnie rescued it from the tide off Garibaldi. It doesn't take the haunted eyes of a war veteran like Lonnie to trace the outline of a horse's head in the cedar grain, and I felt its power in Lonnie's dance. One of America's unsung heroes, Lonnie is hippied-out, as honest as an Oregon blueberry. Who knows what Lonnie has seen riding an incoming tide?

While it travels the ocean currents, driftwood provides homes for seabirds as its flora feeds fish and other aquatic species. Meanwhile, gribbles, shipworms, and ocean-friendly bacteria work to decompose the wood, adding unique character to each branch, or limb, or trunk. Washing up onto beaches, driftwood can pile up to form a skeletal structure on which sand dunes may form. Weathered, sun-bleached, gnarled, and riddled with wormholes, driftwood is one of nature's most eclectic and ubiquitous art forms, most beautiful when left unaltered by human hands. God bless the driftwood crafters, can you please take a hint from Lonnie Rose and let driftwood speak for itself?

Driftwood today is celebrated at fairs in a few beach communities along North America's coastlines. But its beauty is even more revered as natural sculpture in Japan, in Barcelona, Spain, and in other European cities, where major driftwood exhibitions bring together flotsamists from around the globe. I've never personally attended one of these gatherings, but I can just imagine Lonnie showing up in Barcelona and wowing the crowd with his cedar-grain horse.

Perhaps the most important driftwood log to complete a transoceanic voyage was the tree trunk that reputedly crossed

Driftwood connoisseur Lonnie Rose.

the Pacific Ocean from Peru to Polynesia, carrying on its back the exotic sweet potato. As far back as 10,000 BC, potato tubers were harvested in Peru. By 2500 BC, South American natives were cultivating potatoes and creating new varieties. The sweet potato varieties were particularly well suited to hot, moist climates. Centuries before Spanish and Portuguese explorers discovered potatoes in the Americas, the tubers were already being cultivated in New Zealand and in the South Pacific Islands. Paleontological evidence suggests that by AD 400, potato tubers had reached eastern Polynesia, and by AD 1300, sweet potatoes were being cultivated in New Zealand. From Polynesia, the tubers traveled northward into Asia, reaching China and Japan, where they became dietary staples. Although no clear evidence has yet definitively proved how the tubers first arrived in Polynesia, many paleontologists believe the sweet potato most likely traveled across the South Pacific as tubers attached to a driftwood log, or as seed capsules that floated. Remember that at Thanksgiving dinner.

The Chilean Blob

Organic flotsam, as we have come to learn close-up, comes in all species. Los Muermos, Chile, in 2003, received a frightening visitor to its shores. The washed-up thirteen-ton blob of amorphous tissue contained no bones or scales or armor of any sort, and while marine biologists scratched their heads in wonder, some locals speculated the thing from the sea was a giant octopus.

When scientists studied the tissue and found sperm whale DNA and dermal glands, the giant-octopus-monster buzz suffered a blow, though it didn't die out. Then biologists checked some preserved samples of other sea monsters that had washed up in other locales, including the Giant Octopus of St. Augustine, Florida, circa 1896; the Tasmanian West Coast Monster, circa 1960; the Bermudan Blob, circa 1992; the Nantucket Blob, 1996; and finally, Bermuda Blob 2, in 1997.

Test results confirmed all these monstrous sci-fi gooey gobs of flotsam were, in reality, *ballena restos*, or, in French, *vestige baleine*, or, in Swahili, *nyangumi vipinda*; in any case, more than whale yack but less than Moby Dick. Translation: whale remains.

The Drifters

We already know that Cro-Magnon man wore shell jewelry. Now it comes to light that at least for the past thousand years, Native Americans have made necklaces from seedpod flotsam. Especially along tropical coastlines and subtropical beaches, exotic seedpods wash in off the tides. In 1894, *Garden and Forest* magazine published an article by C. H. Coe, a collector of sea beans on Florida's beaches. Although sea beans had for centuries, possibly millennia, been put to sundry uses by humankind, Coe's article, which focused on how to polish sea beans, is the first recorded literature in the United States.

Sea beans are seedpods of tropical plants, woody vines, and trees. The sea bean's exterior pod or outer seed coat is virtually waterproof. Inside, an air pocket, often supporting a fertile kernel, gives the pod buoyancy, making it a great candidate for bobbing along on ocean swells and currents. A legend from the Azores says Columbus, while searching for the West Indies, learned of seedpods washing ashore on the Azores. Specifically, Columbus learned of a heart-shaped, mahogany-colored pod now named *Fava de Colom* and commonly called the sea heart because of its vaguely heartlike shape and its deep mahogany-red shade. Columbus, according to this legend, had seen sea hearts in Europe, where the pods were fashioned into snuff boxes. The pods had washed up onto European beaches off the North Atlantic Current. Some collectors refer to this sea bean as a kidney bean, which begs the anatomical question, "Which is the more accurate description of its shape?" I vote for heart, but then I've never autopsied anything owning a heart and a kidney. The sea heart comes from *Entada gigas*, a woody vine that grows in America's tropical climes. This humble bean had navigated the Atlantic's Gulf Stream before entering the North American Current. Columbus also identified at least one species of bamboo in the West Indies, which he erroneously thought must have come from the East Indies. Although he got the land mass wrong, Columbus nevertheless understood ocean currents, thanks to such arcane clues as the floating adventures of tropical seedpods. If flotsam can help an explorer navigate alien waters, it must be worth something.

In fact, sea beans of many varieties had been traveling ocean currents for thousands of years, journeying both north and south from their native soil. The northward drifters often float as far as Western Europe before beaching on the shores of England, Scotland, Ireland, and Wales; some few travel even farther north to the Netherlands and Norway. Originating in the Amazon rain forests and on the beaches of the American tropics, the seedpods are washed by rains and tides into the Gulf Stream current. Their tough outer coats survive thousand of miles of saltwater drifting, though some with weaker exterior shells sink along the way. Once beached in a northern temperate climate, the beans might sprout, but will soon die from the cold temperatures. Scottish folklore tells the story of the Mary bean, actually a washed up morning glory seed, which was used in Scotland to ensure safe childbirth. The woman delivering held the bean while her midwife chanted a Gaelic verse. In Norway, women fought labor pains by sipping a brew of tea or ale, from a cup made of sea heart skin.

Ed Perry IV, a Florida park ranger, and the late John V. Dennis, a botanist, biologist, ornithologist, and writer, studied sea beans for decades. Perry's initial interest in sea beans and drift seeds came from his grandmother, Helen Wright Risler, who operated the Sea Bean Boutique in Cocoa Beach, Florida. Dennis, who passed away in 2002, and Perry were among the charter members of The Drifters, a seedpod squad that publishes an international sea-bean and drift-seed newsletter appropriately named *The Drifting Seed*. The Drifters soon initiated a dialogue among botanists, oceanographers, serious sea-bean collectors, and interested beachcombers that has developed into the International Sea Bean Symposium, an annual conference held in Florida aimed at discovering everything there is to know about these marvelous bits of flotsam.

And they are marvelous: One glance at the pages of *Sea-Beans from the Tropics*, a book by Perry and Dennis, may hook a beachcomber into collecting these often colorful, sometimes striped or grooved, seedpods. I've seen collectors wearing sea-bean necklaces that would absolutely pop from the pages of *Vogue* magazine, and a revival in sea-bean jewelry hit the summer 2005 fashion runways. Seedpod jewelry is now hawked by street vendors, and my supermarket sells seedpods by the pound to use in flower arranging. Of course, these are harvested at the sourec, not off tide lines.

Atlantic beaches along North and Central America collect thousands of sea beans annually, often found in sargassum weed and floating marine life wrack that the outgoing tides leave behind. Perry and Dennis say nearly two hundred sea-bean species have washed up so far on Florida's beaches, perhaps the world's finest sea-bean collection beaches.

A number of factors affect where and when these tropical gems will wash up. In the Atlantic, the Gulf Stream, the world's strongest current, feeds into the North Atlantic Drift, which carries sea beans from as far south as the Amazon to as far north as Scandinavia. Weather also plays a role; hurricanes in particular can shoot the sea beans off into other surface currents, sending them on a less predictable odyssey than if they remained in the Gulf Stream.

The Drifters of Florida remember fondly the legendary "queen of sea beans," Cathie Katz. Katz, who grew up in Holland, was in her last years a resident of Melbourne Beach, Florida, was founder and editor of *The Drifting Seed*, a founder of the International Sea Bean Symposium, and author of several volumes in the *Florida Nature* series. Katz was known for being the sole beachcomber brave enough and impassioned enough to set out along the beach during a hurricane looking for sea beans and other flotsam. Having combed beaches in Holland, Spain, Portugal, and New Jersey, Katz often said that Florida's beaches far exceeded any of those beaches in flotsam yield.

A side note on Katz's flotsaming: She once found a plastic-wrapped package washed up; inside were identification papers belonging to a Cuban man, including his driver's license and birth certificate. They had washed ashore during the 1994 Cuban exodus. A year later Katz tracked the man down in Miami and returned the papers to him, relieved to learn he had survived his voyage. Katz also found a porcelain Portuguese fishing float the size of a bowling ball—virtually unheard of in today's era of proliferating plastic fish-net floats.

But can exotic seedpods germinate once on foreign soil? Although seedpods venturing far north into Scandinavia and the British Isles cannot germinate in those cooler climates, Ed Perry in Florida has managed to sprout a brown hamburger bean and has grown it to maturity. The plant has already produced little hamburgers of its own. Because sea beans are still one of botany's least-studied subjects, Perry believes Florida's Brevard beaches receive drifting seeds from plants as yet unknown to botanists.

Sea beans also float across the Pacific Ocean from Southeast Asia. They've been found on Pacific Northwest Coast beaches. Some are still able to sprout, although the vines they produce will not withstand frost. Hamburger beans, which resemble Big Macs, have turned up on Washington State beaches. The hamburger bean contains L-dopamine, a natural form of the chemical sometimes used to treat Parkinson's disease.

Velella velella: Flotsam's Harbinger

The first time I consciously encountered *Velella velella*, I fell to my knees on the tide line and gave my heart to its translucent blue sail as the curious jelly tacked in the shallows and then beached on the sand. I did not know then how critically *Velella velella* influences a seasoned flotsamist's harvesting habits. Since then I've learned that when an aggregation of *Velella velella* washes up, flotsam's on its way. But upon my first encounter, all I saw was its incandescent beauty and its ingeneous design. This remarkable gelatinous hydrozoa with its own sail and the ability to tack is commonly called a By-the-Wind Sailor. No wonder I fell in love at first sight.

The bright, transparent navy-blue animal is an enchanting combination of cartilagelike skeleton and gas-filled pockets. Oval-shaped, about four inches in diameter and two inches high at the vertical triangular crest set diagonally across the top, By-the-Wind Sailors ride warm water currents.

Adrift on open temperate seas, By-the-Wind-Sailors are social to the extreme, voyaging in aggregations of tens of thousands of individuals. The jelly itself is comprised of a colony of zooids, including gastrozooids and gonozooids. The gastrozooids feed on fish eggs, shrimp eggs, and planktons, while the gonozooids perform the reproductive duties, ejecting tiny medusa.

When driven ashore en masse, sometimes in mile-long flotillas, they form great mounds of blue gelatinous goo, and—there's always a catch to love at first sight—in death emit a ghastly odor detectable half a mile inland.

Some By-the-Winds tack to the right while others tack leftward. Both types apparently mix together in the central oceans, and I rather doubt that any sort of leftward discrimination exists in their world. But when a strong wind whips up the waters, the jellies tack toward one shore or the other, depending on their bent. Along American Atlantic coastlines, the beached jellies are usually lefties, and strong seasonal winds blow them northwesterly toward shore, while on American Pacific coastlines, By-the-Winds tend to tack to the right as prevailing northerlies keep them offshore, until spring and summer, when strong southerly or westerly winds send them tacking to the beach en masse.

By-the-Winds usually mark the imminent arrival of glass floats and other wonderful things washing in on the tides. That's because *Velella velella* is driven ashore by storms—the same storms that rake the central oceans of their floating debris and send it careening toward landfall.

II. Adrift at Sea

My love, alas, I'm lost at sea
And sinking fast.
In my hand your photograph
Accompanies me in Neptune's bath.

Anonymous

A Brief History of Bottled Romance, Adventure, Research, and Heartbreak

One summer morning on the beach at Green Acres, Maury Island, I was just beginning my scan of the fresh tide line when I heard my brother call my name. I was four years old, and Rob, who was five, ran faster than I, and therefore had already covered most of the washed-up flotsam during our first beachcomb of the day. When he called out, I did a lickety-split and found him crouching down at water's edge, clutching a glass bottle. "Let me see," I said. He responded typically by hiding the bottle from me. If he wasn't going to show it to me, why had he bothered to call me over? Any older brother or younger sister knows why. It wasn't until he'd gone off like a dog with a bone into the driftwood canyons where he knew I would not dare disturb his privacy that I realized he'd found a message in a bottle. That, or a full bottle of Miller High Life. In those days I ran with the maxim that the sea washes up only one great treasure per tide, and today, at least, Rob had scored.

When he emerged from the driftwood, holding the bottle aloft, I knew he'd hit pay dirt. All flotsamists are braggarts and competitive to within an inch of murder. Rob's message in a bottle came from somewhere far away, as I recall, traveling several years before washing up on this beach in Puget Sound. I remember the message included a return address, and my brother wrote and mailed a reply and then received a letter back. That's all I remember about the message in the bottle, except that its author was a grown-up, not very interesting, and therefore forgettable.

It's been said that the sea can destroy great ships but even a hurricane can't stop a bottle with a message inside. Actually, I made that up; it just sounds proverbial. Part science, part romance, the saying holds truth. Because of its floatability, a bottle can travel thousands of miles without sinking. Because of its durability, it withstands the ravages of decades traveling in saltwater. The mere mention of a message in a bottle triggers poignant romantic visions of sailors lost at sea, of shipwreck survivors cast adrift, of lonely mariners trapped in the doldrums. Inevitably, a bottle with a message inside plucked from the sea or off an ocean beach results in someone's heartache or rejoicing. Women who believed their sailor husbands were lost at sea have been startled when the bearer of a bottled message suddenly appears at the front door, the message announcing the long-gone salty dog is alive as a newborn snake.

Rulers of great nations have ordered their naval officers to record their locations and battle plans on notepaper, tuck it inside bottles, and trust the sealed bottles to ocean currents. Often as not the strategy failed; the bottle stoppers leaked or the bottles washed up against remote boulders and smashed to smithereens, the vital information set adrift to eventually decompose or sink. Battles were won or lost depending on messages in bottles. Yet the first recorded use of messages in bottles placed into the sea is a tale of scientific strategy aborning. In 310 BC, Greek philosopher Theophrastus, aiming to prove the Mediterranean Sea was formed by inflow from the Atlantic Ocean, tucked messages inside bottles, which he tossed into the Mediterranean. Historians generally agree Theophrastus received no replies, yet history isn't always accurate, nor thorough. I prefer to think the bottles washed up on distant shores and their finders, unable to read Greek, simply recycled the bottles for olive oil.

Legends abound of lonesome sailors trusting their lovelorn letters to the sea, of stranded mariners on derelict vessels lofting desperate SOS pleas in bottles, and even last-ditch survivors of shipwrecks who jettisoned their own grim epitaphs before their ships went down, reporting the tragic news. In Christopher Columbus's ship log, the fifteenth-century explorer on his return from discovering the New World recounts that one of his three ships, *La Niña*, which was carrying him back to Spain, was caught at sea in a great storm over the Atlantic, and was in danger of sinking. In a brief document, Columbus reported his discovery of this new land and sealed the document inside a small cask, along with a note requesting that the cask's finder deliver the news to Queen Isabella of Spain. He then tossed the cask adrift on the stormy Atlantic. But *La Niña* didn't sink and Columbus returned safely to Spain with his ship's log and tales of the New World. What happened to the message in the little cask is unknown, but the lesson of the tale is that Columbus in the fifteenth century understood the circular pattern of the ocean's currents and knew chances were good that his message would drift eastward, eventually washing up on a European shore.

Reacting to a report that a Dover boatman had found a message in a bottle and had opened the bottle and read the message inside, an appalled Queen Elizabeth I, in the sixteenth century, appointed an Uncorker of Ocean Bottles and made it a capital offense for anyone but government officials to open such a message. The queen hoped the harsh penalty would protect strategic messages in bottles sent by British fleets at war, a military communication practice that continued for centuries. Ignorance was no excuse for breaking the law, and even the Dover boatman's cries of "Who knew?" didn't save him from the gallows.

In 1784, Japanese treasure hunter Chunosuke Matsuyama and his crew of forty-four stranded on a coral reef in the South Pacific. Without fresh water or food, the desperate captain carved an SOS into a piece of wood, sealed the message in a bottle, and lobbed it into the sea. One hundred and fifty years later, in 1935, the bottle was found on a beach of the seaside village where Matsuyama was born. The Japanese treasure hunter and his crew presumably starved to death long before the SOS washed up; still, you can't beat that bottle's dead reckoning.

America's favorite renaissance man, Benjamin Franklin, found a way to speed the mail by experimenting with messages in bottles. Appointed postmaster general for the American colonies in 1737, Franklin noted that the postal department was painfully short on navigation charts. He decided to create an updated and accurate chart for navigating the Atlantic between North America and Europe. At the time, the postal department hired whaler captains to command the colonies' mail ships sailing to England because whalers understood the ocean currents better than other sailors, and so their ships made it to England much faster than the British mail packets heading to America. Franklin, inveterate collector of useful information, had bottles dropped into the Atlantic's Gulf Stream, with messages inside asking the bottles' finders to contact him. Eventually some of the messages were returned. Franklin then combined the whalers' knowledge of currents with the information he gleaned from movement of the bottles to create a chart on Atlantic currents—charts so accurate that they are today essentially unchanged.

In 1861 a message in a bottle was plucked off a beach on the eastern side of Grand Turk Island. The message inside the bottle read, "H.M. Sloop Ringdove 25th November 1859 Lat 26.21 Long 18.7 by observation. This paper was thrown overboard at noon on the above day having just entered the North East trades. Force wind 3, along Northwesterly swell. Barometer 30.43, Thermometer 75, seawater 73. R. G. Cragie Commander." An accompanying note asked the finder to forward the information to the British Admiralty, suggesting that the British Navy continued using the Elizabethan method of messages in bottles to convey official information.

In 1875 the captain of the Canadian barque *Lennie* and his officers were murdered by mutinous sailors. Thinking ahead, the mutineers spared the steward, who was able to navigate. The steward then steered the *Lennie* toward France, telling his captors that they were headed for Spain. Meanwhile, the steward furtively lobbed several SOS messages in bottles, including a description of the *Lennie* and its calculated bearings. One of the bottles was picked up and delivered to French authorities, who sent a rescue vessel. The mutinous sailors were arrested and the steward told his story.

On April 20, 1894, Ralph Rogers, a passenger aboard the *Marion Lightbody*, became homesick for Scotland. As the ship sailed off Cape Verde, Rogers decided to write a letter to his sister, Marion, and send it to her via the Neptune's post. Rogers wrote, "Ship Marion Lightbody, Lat 25 degrees 2' N, Long 23 degrees 15' W, 20th April 1894, All well on board, R C Roger," with the postscript, "Finder please send this to Miss M Rogers, Lancaster Terrace, Kelvinside Glasgow Scotland."

On January 16, 1896, in Glasgow, Marion Rogers received her brother's handwritten letter along with a note: "Grand Turk 29th Dec 1895 Miss M Rogers, As requested by the writer that the person that finds this would please forward it to you. I have undertaken to do so and hope you will receive it quite safe. I found it on the 28th Dec. Yours respectfully, Albert T Wynns, Grand Turk, Turks Island, West Indies."

One of Glasgow's newspapers covered Marion Rogers receiving her brother's message from Wynns, the braggy reporter commenting, "I can see from the map that the bottle fell within the 'sphere of influence' of the North Atlantic current, which after flowing southwards along the coast of Africa for hundreds of miles turns sharply to the west at Cape Verde, sweeps across the Atlantic and spends itself amongst the northern West India islands of which Grand Turk may be reckoned as the *avant garde*. With the chart of

No message, but seal intact on this washed up bottle of Chateauneuf-du-pape. Vintage uncertain.

oceanic currents before my eyes, I might have predicted the destiny of this bottle, after a course of three thousand miles, with almost absolute exactness. It took 619 days to accomplish the passage, so it would appear that Neptune's post is somewhat slow. . . ."

Toward the *fin de siecle*, Matthew Fontaine Maury imitated Ben Franklin's messages-in-bottles strategy to create U.S. Navy navigational charts, and the pre-casino Prince Albert of Monaco dropped bottled queries into the Atlantic in an effort to study Gulf Stream currents as they approached Europe. The method resulted in the prince's discovery that the Gulf Stream divides in the northeastern Atlantic, with one current, the North Atlantic drift, branching toward Great Britain and Ireland while the other current, the Azores Current, runs south along Spain and Africa before heading west again. Albert's discovery aided the Allies during World War I. Using his splitting Gulf Stream theory, military officials were able to predict the travel patterns and landing sites of explosive mines the enemy had set adrift, and thus locate and disarm them before they exploded.

Marine science still employs the message-in-bottle strategy. Combining the use of drift cards—floating cards that are more practical than bottles—with satellite imaging and computer technology, oceanographers are able to predict with considerable accuracy where and when an object thrust into the ocean will wash up. But nature often intervenes, surprising scientists with sudden changes in wind direction, rogue waves, or other natural trickery that sends drift cards headed for, say, West Africa to end up in Nicaragua, or even to disappear from satellite tracking altogether, perhaps showing up one sunny winter afternoon off the coast of Sicily. Stranger things have occurred. Take, for example, the legendary Flying Dutchman.

Launched in 1929 into the southern Indian Ocean by a German marine science expedition, a bottle that was nicknamed Flying Dutchman contained a message that could be read without breaking the bottle. The finder was asked to report, via ground postal services, where the bottle was found and then toss it back into the sea. Because travel time depends on winds and currents, such a bottle might languish in a becalmed sea or it might catch a swift-moving current like the North Atlantic Gulf Stream, where it could drift up to four knots, possibly covering a hundred miles in a single day.

The Flying Dutchman soon slipped into an eastward current that carried it to the tip of South America, where it was found and reported and then lobbed back into the ocean numerous times. Eventually the Flying Dutchman made it back into the Indian Ocean and passed near the spot where it had originally been jettisoned. Traveling still longer, it finally came ashore on Australia's west coast in 1935. When it landed for the last time, the Flying Dutchman, according to calculations based on its finders' reports, had traveled sixteen thousand miles in 2,447 days (slightly more than six and a half years), traveling approximately six nautical miles a day.

Science rules, but heart holds court. In 1914, during World War I, Thomas Hughes, a British infantryman, missed his wife while on a ship crossing the English Channel. Hughes wrote her a letter, sealed it in an empty ginger-beer bottle, and tossed the bottle overboard. Two days later,

Hughes died in battle. In 1999 a fisherman discovered the bottle in an estuary of the Thames. The fisherman located Hughes's daughter, who was then eighty-six years old and living in Auckland, New Zealand. He flew to Auckland to personally deliver the message.

A letter written on the decks of the torpedoed and sinking *Lusitania* in May 1915 was later discovered, its final sentences, "Still on deck with a few people. The last boats have left. We are sinking fast. Some men near me are praying with a priest. The end is near. Maybe this note will . . ."

In 1944, on a beach in Maine, a bottle was found with this message inside: "Our ship is sinking. SOS didn't do any good. Think it's the end. Maybe this message will get to the U.S. some day." The message was traced to the USS *Beatty*, a destroyer torpedoed off the coast of Gibraltar on November 6, 1943.

In 1953 a message in a bottle was discovered washed up on a Tasmanian beach. It had been lobbed into the sea by two Australian soldiers aboard a troop vessel carrying them to war in France. When a picture of the message was published, a woman recognized the handwriting as that of her son, who had been killed in action in 1918.

The 1976 film *Voyage of the Damned*, based on the book by Gordon Thomas and Max Morgan Witts, reprised the horror of some 850 Jewish refugees fleeing Nazi Germany aboard the SS *St. Louis*, bound for Cuba. Cuban president Frederico Laredo Bru, who had guaranteed the *St. Louis* passengers asylum, changed his mind. As the *St. Louis* languished in Havana harbor from May 27 to June 6, 1939, waiting for negotiations to decide the issue, the passengers formed a committee to represent them in their pleas for sanctuary. When President Bru refused to accept the refugees, the passengers tossed scores of bottled pleas for help into the sea. The *St. Louis* headed back to Europe, where it unloaded its desperate passengers in Belgium, Holland, France, and England. In 2003, at a book sale in Bath, England, a John Moore discovered an edition of *Voyage of the Damned*. Inside he found a piece of notepaper with a handwritten message, "Please help us President Bru or we will be lost." It had indeed come from the *St. Louis*, one of perhaps hundreds of messages in bottles the stranded refugees tossed overboard.

In an old-time version of Match.com the lovelorn trusted Neptune's post with their heart's desires. In 1956 while at sea with nothing very interesting to occupy his mind, Swedish sailor Ake Viking jettisoned a long shot overboard. The message inside Viking's bottle asked for any pretty girl who might find the bottle to write to him. Months later a Sicilian fisherman plucked Viking's bottle from the sea. After reading the note, the fisherman, as a joke, gave it to his daughter, Paolina. Paolina wrote to the Swedish sailor and a correspondence ensued. Eventually Ake Viking visited Paolina in Sicily. In autumn 1958, Ake and Paolina married.

After World War II, in 1946, the U.S. Navy used drift bottles to map areas where Japanese mines might wash ashore following storms at sea. By this time oceanographers had begun employing drift bottles in research on currents affecting navigation, fishing, derelict vessels, and marine environments. Information gleaned from drift studies have saved the lives of sailors and swimmers lost at sea. Missing cargo ships and pleasure vessels have been located based on drift studies, and the trajectory of major oil spills plotted. Municipalities along river routes use drift bottles to determine the course their trash takes as it heads downriver. Some modern research projects involve several thousand drift bottles, or drift cards, placed into the ocean, and often the scientists will offer a reward for reporting where and when one is discovered. A knowledge of the sea's surface drifts and currents helps predict when and where flotsam will wash ashore, thereby substituting scientific accuracy for the mystery and romance that once accompanied the washed-up message in a bottle.

Yet even today a certain excitement accompanies the discovery of a research drift card or bottle, the reward being less monetary than the satisfaction the finder feels knowing he or she has contributed to a scientific research project. Personally, I'd rather find a love note, or an SOS.

Coastal schoolchildren around the globe often drop bottles in their communities' bays and estuaries hoping for a reply from far away. Since its founding in 1998, the Kuroshio Monogatari Cheerful Children Association, in Genkinakonokai, Japan, has encouraged international pen pal friendships among children through the release of messages in bottles into the Pacific Ocean via the Kuroshio Current off Japan. The zipper manufacturer YKK provides bottles with water-resistant zippers for the project. The zippers keep water out and air in so the bottles won't easily sink.

Schoolchildren in the United States are also fond of trying this marine experiment. Once in a while a kid gets really lucky. In 1980, at Providence, Rhode Island, Pleasant View

School teacher James Westerman encouraged his students to try sending messages in bottles. The students prepared their messages, stuffed them into bottles, and tossed them in Narragansett Bay. In November 1982, one of Westerman's students, Frank Marston, received a note from a Spanish merchant naval officer in the Canary Islands off North Africa. But that's not the end of Westerman's tale: In 1984, young Nomp Travis, another of Westerman's bottle-lobbing students, received a reply from ten-year-old Jayne Ayre (no kidding) of Barnstaple, England. Jayne Ayre wrote that she had come upon the bottle while strolling the strand with her father on January 29, 1983. "I found your name on it," wrote Jayne Ayre, "and was thrilled to see it had come all the way from America." With her reply she enclosed a clipping from the North Devon *Advertiser*, an article reporting her flotsam discovery.

The Emperor of Flotsam

Oceanographer and university professor Tadashi Ishii is the undisputed global Emperor of Flotsam. A professor of oceanography at Kyushu Industrial University, Ishii is president of the Flotsam Scientific Society in Japan and author of *Shinpen Hyouchakubutsu Gitenn*, or, *The New Encyclopedia of Flotsam*, its second edition published in December 2002. *Hyouryubutsu*—Japanese for flotsam—is Ishii's passion, and he's found some remarkable objects on Japan's beaches. Among flotsamist Ishii's message-in-bottle collection is a bottle with a strong scent of perfume. The note inside, apparently in a schoolgirl's Japanese script, said, "I wish my family will be happy forever and my grandma live for a long time, that I'll have a wonderful boyfriend. I hope my mama is happy in heaven, I want to be a nurse. I wish happiness for my friends and my brothers. . . . This is for the god of the sea."

In June 1979, Ishii pulled a whiskey bottle off the tide. Inside was a note from a lovelorn and spurned schoolboy who presumably consumed the bottle's contents before penning a message to diverse sea gods and placing his heart in their fickle hands. That same year, in September, Ishii found a bottle with a pencil inside and a note that said, "Could you be my friend?" The girl explained that she was very ill and had been in the hospital for a long time. Dr. Ishii, the scientist, possesses a special passion for bottled romance, and, of course, contacted the ailing child.

Even fishermen get the blues, according to Ishii. He found a bottled message in Japanese script on the Ashiya seashore, which he believes was written by a fisherman. The note said, "I wish next year will be a great year for me." Hey, buddy, don't we all?

In August 1991, on Okinawa Beach, Ishii found a champagne bottle containing a sheet of notepaper and an American dollar bill. The paper wouldn't yield to Ishii's attempts to remove it, so he broke the bottle. The message, written in English, read, "February 19, 1989, Southern California. Sixty miles off the California coast, this bottle was dropped from the air." The person who dropped the bottle into the Pacific Ocean wanted the bottle back, and so included the dollar bill as postage, apparently having miscalculated the dollar-to-yen exchange rate.

Ishii packaged up the broken bottle shards and mailed them, along with photos of the intact bottle, to the sender, asking him, "How many bottles are you guys dropping into the water?" In early December, Ishii received a response: "We began in 1988, dropping the bottles from off the Southern California coast. Most of bottles were found within a 20 km area. Every February through April, I charter an airplane from Santa Barbara and fly westward to Point Conception over the Pacific Ocean dropping over 100 bottles gradually. So far, this year, I've dropped 355 bottles. Different types of bottles. 14 bottles were found in Philippines. Two bottles in Hawaii. Also Rio de Janeiro, and Columbia. One was found in Japan, Hiroshima Prefecture—dropped on March 26,1988, from San Nicolas Island, it was found in Japan on October 8, 1990. I use some polyethylene bottles because they usually float, but even so, to strengthen the bottle, I put it on a rooftop for 2 years," to cure so it wouldn't break.

When Ishii found the bottle, the paper wasn't dry. "The American guy said I was the first person whose paper wasn't dry even though he put red rubber sealer on the bottle and plastic over that."

Politics Afloat

During the winter of 1994–95, a series of brutal storms on Boswell Bay in Hinchinbrook, Alaska, sent waves clawing at the already eroding sand and gravel Strawberry Beach. The erosion laid bare sediment that hadn't seen daylight for decades. Beachcombers Brooke and Gayle Adkinson noticed an amber-colored bottle partially buried in the scarp. The bottle had a note inside. Water had seeped into the bottle and the paper was soaked. Using a pair of tweezers, the couple was able to retrieve and unfold the paper, but it came apart. After three hours of tedious work, the Adkinsons had pieced together the note, and to their astonishment they saw it was composed in three languages: Russian, Japanese, and English. One side of the document included the following information: "Imperial Russian . . . thrown 5/ 18 July 1913, N 54°26′, E 141°55′." The other side of the document included the words, "Vladivostok . . . East Siberia . . . The Pacific Hydrographical Expedition."

If the information was correct, and the notepaper and bottle did date from Czarist Russia, the bottle had been tossed overboard from an icebreaker in the Sea of Okhotsk a bit north of Sakhalin Island during a 1913 exploratory mission conducted by the Imperial Russian Navy in efforts to find a shorter sea route to Japan. The czar's hydrographical expedition had tossed bottles with messages inside hoping for responses from their finders, information they would use to help chart ocean currents along Siberia's Arctic Ocean coastline. At least one of their bottles got loose on the currents and after traveling southward to the Japan coast, the Adkinson bottle headed east on Pacific Ocean currents before being sucked into the Gulf of Alaska current from whence it found rest for a time at Strawberry Beach.

Tadashi Ishii's *Encyclopedia of Flotsam* references messages in bottles produced in recent decades by Taiwanese political activists who printed propaganda and Taiwanese flags on vinyl sheets, placed them in various colors and shapes of plastic containers, and set them adrift toward mainland China in hopes of encouraging their finders to rebel against the Chinese Communist regime. Ishii describes the canisters as being green, yellow, or blue, with orange or red lids, the lids embossed with the Chinese character for "good fortune" and on the canister bottom, a plum flower symbol. They were produced in eleven different shapes, including jugs, balls, and flasks. The majority of the canisters, says Ishii, were released en masse from naval vessels near China's Fujian Province coastline. According to a Chinese yearbook, over 100,000 canisters had been released by 1983. The tactic continued at least until 1991. Not only written propaganda was released in the canisters; according to Ishii, many contained consumer goods such as perfume, face creams, laundry detergent, and sewing needles. Ishii reports that between 1979 and 1986, sixty-two canisters were found on Japan's beaches. Similar canisters have washed up along the coastlines of Alaska, Hawaii, and Canada.

The Taiwanese aren't alone in creating flotsam propaganda. In February 1997, beachcombers in Niigata, Japan, came upon South Korean canisters that had been aimed at North Korea's coastline. These canisters contained leaflets, canned food, and photographs.

On June 10, 1990, University of Washington oceanographer Richard Strickland, while kayaking in Barkley Sound just off Vancouver Island, Canada, retrieved a bottle he had spied on shore. Peering inside, he saw some papers covered with Chinese characters. But circumstance intervened, and Strickland didn't open the bottle until December 1991. Inside he found six leaflets. One of them was a written appeal for the release of Wei Jingsheng, a Chinese dissident well known in the West. Further inquiry revealed that the bottle Strickland found was probably released in the summer of 1980, when ocean currents and winds proved favorable for sending flotsam mail to the China mainland. The bottle was apparently one of thousands released off China's Quemoy and Matsu islands by Taiwan activists trying to notify citizens in the Peoples Republic of China that Wei Jingsheng had been arrested and detained.

Now here's where the mystery deepens. Oceanographer Curtis C. Ebbesmeyer, at the time working at the University of Washington, became involved and, with several other scientists, attempted to track the bottle's path from the China coast across the Pacific to Vancouver Island. The team agreed that the bottle wasn't sent before October 1979, the month Wei Jingsheng was arrested. They allowed time for the plan to be carried out: Messages had to be written and printed, bottles stuffed and lidded, and, too, there was travel and, no pun intended, execution time. The next summer's favorable winds would have blown toward the Chinese mainland, thereby reducing the number of bottles left in the ocean. Conclusion: The bottle was jettisoned during the summer of 1980.

Next came computer technology, via a program called Ocean Surface Currents Simulation, a revolutionary tool developed by oceanographer W. James Ingraham Jr., another University of Washington researcher. But at the time, the technology couldn't simulate ocean currents along the coast of mainland China, so the oceanographers had to conjecture, based on Ingraham's and Ebbesmeyer's considerable experience, about the movement of the bottle until it reached open ocean. If, as seemed possible, the bottle had entered the Kuroshio Current off the Japan coast and traveled the current north by northeast, it might have reached

Vancouver Island within two to three years. Once off Vancouver Island, the bottle could have lollygagged around Barkley Sound until 1990, when Strickland retrieved it.

Another theory presented reasonable possibilities. The bottle might have entered the Kuroshio Current, followed the North Pacific Drift, and bobbed into the southbound California Drift, then traveled even further south into the North Equatorial Drift, then back into the Kuroshio. It could have made this panoceanic circle more than once before it finally broke out of the currents and washed up in Barkley Sound. The scientists figured travel time during that decade at around four and a half years for one complete circuit of the North Pacific Ocean, so the bottle may have ridden ocean currents for two or more rotations. Although no absolute conclusion could be reached from this limited evidence, the exercise is the sort of detective work oceanographer's relish. But the discovery of a single bottle wouldn't advance science much. If, on the other hand, hundreds of bottles had washed into Barkley Sound, they might represent a shifting drift or current, help predict a seasonal shift, or verify that a freak of nature had occurred. Oceanographers would jump on the new challenge.

Trolling for God: How to Make a Missionary Bottle

In the summer of 1980, at Okinawa, Tadashi Ishii found a beer bottle with a Japanese religious tract inside. Ishii says that some religious Japanese lob holy flyers into the ocean (inside sake bottles?) and that many of them reach Hawaii. In 2004, in Charlottetown, Prince Edward Island, Canada,

Harold Kemp decided to proselytize for his religious faith by bottling evangelical messages and sending them adrift in the Atlantic. Retired at sixty-seven, Kemp said he'd been looking for a pastime when he struck upon the idea of marine evangelizing. Within a few months Kemp had set adrift more than three hundred epistles in bottles. After complaints that he was littering the marine environment, Kemp saw the light, repented, and ceased lobbing Bible verses into the sea.

A Web site authored by Kraig Josiah Rice offers instructions for making "missionary bottles": "Do you want to make and float your own missionary bottles for the Lord? Here is a complete list of directions, where, if followed carefully, you can become a very successful literature evangelist in regards to missionary bottle evangelism work for the Lord. . . . It is important to have a burden for the lost and be continually in intercessory prayer every step of the way in this kind of ministry. You are engaging in spiritual warfare. . . . I hope you and I are up to accepting this challenge."

In his directions, Rice erroneously calls plastic containers biodegradable. He also cautions, "If the bottle has a neck that is too tight or too loose it should not be used. Some modern ketchup bottles have necks that are too wide—they will not hold a tight enough cork and seawater will seep in as a result and rot the Word of God."

SOS

In May 2005, eighty-eight refugees from Peru and Ecuador who were stranded at sea for three days off Costa Rica were rescued from their sinking vessel because the women aboard had insisted on placing an SOS note in a bottle and attaching it to a fishing line they saw in the water. The fishermen found the bottled SOS and reported it to authorities, and the vessel *MarViva* rescued the troubled craft. Although they were saved from drowning, the refugees, who had been abandoned at sea by human traffickers, were refused entry into the United States and were returned to the countries they had hoped to leave behind.

Neptune's Lost and Found

From *The Week* magazine of September 9, 2005, comes a flotsam tale with a happy ending. In the summer of 1966, sailor James Lubeck, while kneeling on his sloop, accidentally dropped his wallet into the waters off Marblehead, Massachusetts. Lubeck's wallet contained cash, three hundred-dollar checks, and ten—yes, ten—credit cards. Thirty-nine years later, twelve miles off the coast of Gloucester, Antonino Randazzo, a commercial fisherman, netted the plastic wallet insert with the credit cards. Randazzo took the time to track down Lubeck and returned the credit cards. Fortunately for Lubeck, fish don't steal identities.

On April 27, 1985, the Associated Press reported the arrival in Los Angeles of a Vietnamese refugee family to a "tearful welcome from an American couple whose bottled message floated 9,000 miles to the shores of Thailand."

Dorothy Peckham and her husband, John Henry Peckham, had offered in that bottled message to sponsor a Vietnamese family's immigration to the United States. Hoa Van Nguyen, a former soldier in the Vietnamese Army, was walking on a beach in Thailand when, he says, a "sixth sense" told him to pick up a bottle he had spied on the tide line. The message Nguyen found inside was the Peckhams' offer to sponsor immigration to the United States. Not long afterward, Nguyen arrived in Los Angeles with his wife, their sixteen-month-old baby, and Nguyen's younger brother. At their first meeting, Nguyen presented the Peckhams, who had worked through a relief program, with a picture he had made while in a refugee camp in Thailand.

In early September 2005, as floods from Hurricane Katrina still choked a devastated New Orleans, Curt Belton, a Department of Fish and Wildlife agent, reported that rescuers in a boat found a wine bottle adrift on floodwaters along Canal Street in the city's central business district. Inside the bottle was this message: "To Whom it may concern: Please send with immediately, ice cold chest of Coors Light. I'm out at this time. Down to wine. Some shrimp and oysters would also be appreciated. Thank you."

When rescuers located the sender, he was relaxing on his front porch with a bottle of wine. He told them he had enough wine for "quite a few days." He wasn't going anywhere.

Love Gone Adrift

In December 2004, Mary Fulton of Rockaway Beach, Oregon, set out for a walk with her two dogs, heading for nearby Nedonna Beach. Nedonna is a dune-covered beach frequently lashed by major Pacific storms. A storm had just passed and Mary kept high to the dunes to avoid the unstable wet sand. Keeping her head bowed against the wind, she spied something poking out of the sand. It looked like a brown bottle. She wondered if it might be a very old bottle, maybe one of those mouth-blown bottles, a collector's item.

"I saw this brown bottle half poking out of the dunes, and then I saw it had a cork," Mary told me. "I picked it up, thinking maybe it was an old bottle, but it had a seam in it, so it wasn't very old. There was something inside, looked like paper. The cork on the bottle wasn't unusual; it didn't look old.

"There was something else inside the bottle besides the paper. I thought maybe it was sand. I opened the bottle."

Mary read me the note from inside the bottle. It said, "These things were given to me by my lover+friend. I cannot truly return to my husband 100 percent if I keep them. Since you have found them they are yours. The diamonds are real. Enjoy them like I can't."

Along with the note was a pair of earrings, a ring, and a tennis bracelet. "The earrings were just simple heart-shaped, with a small diamond in them," she said. "The ring was real soft, like the metal wasn't worth much. . . . The tennis bracelet is gold with what my friend tells me are diamonds. I know this lady who has a tennis bracelet just like it, and she said she believes it's real diamonds."

I asked Mary why she hadn't had it appraised.

"I'm not a diamond person, for one thing," she said. "And I'm not a tennis bracelet person. I guess I'll do that someday. But for me, the excitement was finding the bottle with a note in it."

The note Mary found included a list:

Dancing

Fall leaves—N. E.

Disney World

Europe—(Great Britain)

Caribbean

Mt. Rushmore

Apt. Lake Forest

Sears Tower

CU do corike (Could this mean "See you do karaoke"?)

First snow.

Messenger to a postmodern world, Mary Fulton, accidental flotsamist, is content knowing she helped a tortured love triangle complete its tragic journey. Love delivered to the sea may go unrequited, yet a message in a bottle, awash in passion, returned to land, offers the world's lovelorn a shred of kindness.

III. Flotsam's Evolution

We have lingered in the chambers of the sea
By sea-girls wreathed with seaweed red and brown
Till human voices wake us, and we drown

T. S. Eliot
The Love Song of J. Alfred Prufrock

So I said to my psychiatrist, "I'm not sleeping nights. This elusive floating stone keeps me awake. Why did I leave it on the beach to wash out on the tide?"

The doctor didn't want to hear about floating stones. I pleaded, desperate to talk about it. He suggested I go to a rock person.

"I don't know any rock persons," I said. "Why won't you discuss it with me?"

"Because," said the shrink, "you shouldn't waste your money. My fees are high. Let's talk about what's important to your mental health."

"But this floating stone is critical to my mental stability," I said. "It's bad enough that I didn't keep it, only took its picture. Now I'm obsessed with its image. I can't stop looking at it. And whenever I bring it up in public, people just stare right through me, like you're doing now."

"You're obsessed with images in general," said the shrink. "And most of them are figments of your imagination."

"I am not imagining the floating stone," I said. I reached into my purse, fished out a picture of the stone and handed it to the doc. He studied the picture for all of five seconds—maybe ten—then flicked it back to me. Stared some more, silently.

"Well," I said. "Now do you see it's real?"

"It's displacement," he said. "The stone represents your mother's womb. You're displacing your anger at her for banishing you from that secure existence afloat in the uterine sea."

"Did you say 'displacement?'"

"Transference, I meant transference."

"But that's spot on," I said. "Displacement is the theory that explains how objects like my stone float on water. You see, when an object weighs . . ."

"Time's up, Popeye."

Lagan Loot

Remember Scrooge McDuck? His fabulous treasure store? The Spanish chests overflowing with diamonds, pearls, golden goblets, gold and silver coins stacked to the rafters? Remember the pirates kidnapping little Huey, Dewey, and Louie, holding them hostage and demanding Scrooge's treasure for the release of his nephews? And did Uncle Scrooge relinquish his pieces of eight to buy his nephews' freedom?

The comics of yesteryear taught generations of America's youth, including me, about buried treasure along its coastlines. Donald was my preferred duck, but Uncle Scrooge had the dinero, so I indulged in the old moneybag's adventures, too. My favorite characters were those swarthy-skinned, stubble-faced, handkerchief-head pirates who perennially—often inexplicably in a landlocked urban setting—raided Scrooge McDuck's treasure vaults. But where did it all come from? Sunken treasure, of course, as indicated by the exotic paraphernalia tucked among the loot. Byzantine crosses, Celtic crosses, Maltese crosses—I think the animator had a cross fetish—great teetering stacks of pieces of eight straight off the old galleon trade, chinoiserie, and all manner of exotica as fabulously illustrated by a Hollywood animator whose longest journey had probably taken him once to Bakersfield. What all this Scrooge material did for me was excite my flotsam genes, sending me straight down to the tide line to search for the first sign of buried treasure—a piece-of-eight silver coin.

The idea that you could beachcomb yourself to riches made sense to young dreamers in the land of free enterprise, a continent ringed with unimaginable sunken treasure whose clues wash up along its beaches often enough to sustain the dream even today. Silver and gold coins from shipwrecked galleons have for centuries washed up along the coastlines of North, Central, and South America, the greater numbers found on beaches in Florida and the Caribbean. Scrooge's treasure was real, after all, and as salvage technology improves, more lagan is pulled from coastal waters, proving at least some of the legends of sunken treasure are real.

But the closest I got to that New Spain silver and gold was in the sixties in Mexico City, where I lived and studied (or pretended to) for a time, and on my frequent excursions to the Mexican west coast. In Mexico City I nearly overdosed on silver and gold rococo Guadalupe shrines and gained an appreciation for Mexico's ultimate victory over the Spanish. If this much gold and silver remains in Mexico, I wondered, how much more was looted and carried away on Spanish galleons? The next question followed logically: Where's the stuff that didn't make it to Spain? And so a trip to the coast seemed in order.

In Acapulco I learned more about hammocks and piña coladas than I learned about sunken treasure, at a museum dedicated to Spanish galleon wrecks, where shades of Scrooge McDuck's pirate nemeses haunted me as I viewed real treasure chests overflowing with real pieces of eight and golden goblets and, yes, even crosses and pearls. It didn't occur to me at the time—this was during my peyote era—that much of what I was viewing wasn't Spanish, nor in any way European in origin, but Asian. Fast forward to 1974, which was the year I first visited China.

As a China watcher for thirty-two years, frequently on site, I've come to understand what all that Chinese silk and porcelain are doing in the Acapulco Maritime Museum. This is lagan of the first order, the stuff every Scrooge McDuck fan ever dreamed of, and it truly did come from sunken ships lying within a beachcomber's reach along the coastlines of both American continents. Much of it still lies beneath the seas, too deep and too far out at sea to recover. Most of it, though, forms a rich girdle that hugs the coast close enough to shore to cause a shipload of deep-sea entrepreneurs to sweat bullets competing for the lost lagan of long ago, which may have been the title of a Scrooge McDuck adventure.

Treasure Runners: Shaking Lagan Loose on the Tides

For centuries, beachcombers have found gold coins washed up on beaches. Americans are particularly lucky; more gold and silver coins are buried in the ocean sand on North and South American coastlines than anywhere else in the world. And they keep washing up: All it takes is one good hurricane, a little luck, and being first on the beach after the storm. Serious flotsamists know that the best time to search a strandline is during a storm and, like the mail carrier, weather doesn't deter them.

While Spanish conquerors were looting silver and gold in New Spain, back home, in the mid-1550s, the shipbuilder Alvaro de Bazan de Sevilla had perfected the design of a new type of warship, the galleon. The fabulous new ships were perfectly suited to transport New World riches back to Europe. Alas, their eventual downfall was partly due to a policy no sane mariner would have concocted: A Spanish galleon's captain was not the ship's final authority; he answered to a naval officer who may have had no major seagoing training or experience and to other ship's officers. In what must have mimicked a Monty Python moment when danger arose—threatening weather or attacking pirates or privateers—the captain was subject to rule-by-committee, often resulting in fistfights, bloodied noses, broken jaws, and even murder and mutiny. But until the rule by consensus debacle eventually proved disastrous for Spain and its fleets, much of the New World's riches rode Spanish galleons across the high seas and into the arms of Spanish nobility.

The galleons carried both treasure and heavy armament and proved strong defenders of the king's treasure during times of territorial wars, often fought on the high seas, and increasing skirmishes with pirates. Formidable when encountered, the Spanish fleets soon became the stuff of legend, as much for its unimaginably rich cargo as for the galleons' threatening presence on the horizon. And those sunken treasures that went down with their ships: Over the centuries violent storms have churned up the ocean beds, sending artifacts from those shipwrecks adrift on the tides, some of the lost treasure eventually washing ashore. Time, tides, and shifting sands have alternately buried and unburied the artifacts. Flotsamists know that every beach holds potential treasure, but some coastal waters, like those off Coin Beach, Delaware, are richer than others.

One autumn day in 1979, at Delaware's Coin Beach, a very determined fourteen-year-old youth named Dale Clifton Jr.—no doubt a Scrooge McDuck comics reader—stood on the strandline and vowed to find a coin washed up from an old shipwreck. If you're searching for a coin, certainly Coin Beach is the place to start.

"Took me one year and two months," says Dale Clifton today. "Then one blustery December day, I reached down into the sand and plucked up a 1785 King George. I've been salvaging from shipwrecks ever since."

The 1785 King George III halfpenny that fourteen-year-old Dale Clifton found ignited a passion in the young man that has burned for twenty-five years and shows no signs of cooling. Clifton taught himself to scuba dive. He became an avid beachcomber. Soon he had collected enough shipwreck artifacts to open a museum, and so he did. Since the first coin, Clifton has salvaged Spanish and Inca gold, gold bars, goblets, jewelry, silverware, porcelain dishes, weapons, and more than 200,000 coins from dozens of shipwrecks off Delaware and Maryland. Working with dive crews, he has located and painstakingly recorded details of shipwrecks, including their crews and passengers, as well as reclaimed the victims' personal property.

Scouring beaches, collecting flotsam, often using metal detectors, and diving to shipwrecks both in Delaware Bay and off the Atlantic shore, Clifton has made sure the shipwreck treasures are preserved for future generations. With colleagues from the Delaware Discover-Sea Shipwreck Museum, which he founded, the undersea detective has also conducted controlled archaeological digs along the shoreline, discovering how colonial American beach folk lived and related to the sea. The Discover-Sea Shipwreck Museum on Fenwick Island, Delaware, one of the world's few lagan museums, houses thousands of articles of salvage from these shipwrecks and beach digs, and is so historically rich and hip it has been the subject of a Discovery Channel documentary. Personal items of victims lost at sea triggered in me a sense of grieving over people I'd never met. Seeing their clothing, jewelry, personal journals, and such intimate items as grooming tools brings them back to life. These artifacts, Clifton says, remind him that the most important treasures salvaged from sunken ships are the memories of passengers, often entire families, who died heartbreakingly close to shores they had dreamed of calling home.

Some estimates put the number of shipwrecks off the eastern Florida coast alone as high as three thousand ships, several hundred of those being Spanish treasure ships, many partially or completely buried under sand. Discoveries of shipwreck flotsam are no longer common along Florida's beaches, but Web sites devoted to salvaging buried treasure report occasional finds of a gold coin, or a gold filigree Spanish cross pendant, plucked out of Florida's white sands.

Not only Spain lost ships along the U.S. Atlantic Coast; over the centuries, vessels of British, Dutch, Russian, and other national origin, carrying passengers and rich payloads, have gone down in heavy weather within a few miles of the coastline, their sea graves never discovered. One of Dale Clifton's ongoing searches is for the *Juno*, a ship reportedly carrying twenty-three tons of silver when it went down along Delaware's Delmarva Coast.

On October 1, 1802, the *Juno* departed San Juan, Puerto Rico, with nearly a thousand passengers. A week later she encountered rough seas and foul weather at 33 degrees north latitude. Against a stiff nor'easterly she struggled northward along the East Coast, and on the morning of October 24, the crew was relieved to spot an American schooner, the *Favorite*, which responded to the *Juno*'s distress flags. Together for the next two days, the ships struggled up the coast. Then, on the night of October 27, the weather turned foul and the *Favorite* lost sight of the *Juno*.The *Juno* was never seen again, nor has her silver bounty been recovered. But with new diving and salvaging technology, Clifton believes the ship may eventually be located and its treasure recovered.

Twenty-three *tons* of silver. Even platinum snobs can appreciate this picture.

In 1565, Spanish shipyards in the Philippines began building the legendary galleons, only these vessels were outfitted for trade between Manila and New Spain, the eastbound galleons crossing the Pacific to make port at what is now Acapulco. Maintaining the European galleon design, the Manila galleons were built of Asian hardwoods like teak and mahogany, and planking was *lanang* wood, with a surface hard enough to withstand a cannonball shot. Beachcombers for centuries have been finding pieces of lanang and teak along North America's West Coast beaches, and residents often speculate about the Spanish treasure ships built from those woods.

Manila to Acapulco was, at that time, the longest navigation distance in the world. These nine-thousand-mile sailings generally took four to seven months. Besides importing Asia's finest products, the fleets also delivered Chinese merchants and Filipino sailors to the Americas. Ship captains understood the North Pacific's surface currents and knew how to take advantage of the trade winds. Still, crossing the Pacific Ocean, even on a mighty galleon, was a life-or-death roll of the dice, one of the greatest risks a person ever encountered. Even the most privileged passengers endured hardship and unsanitary conditions in confined spaces on ships that often carried as many as a thousand passengers. One galleon's log from a crossing between New Spain and Manila notes that a Spanish noblewoman, reacting to conditions aboard ship, went mad and jumped overboard to her death. Meanwhile, pirates, privateers, and enemy navies lurked in the ocean along with the spectre of deadly hurricanes and typhoons, disease, and starvation. Sorry to say, rats, too, though not listed on a ship's manifesto, rode the high seas carrying deadly cargos of germs and disease.

The Manila fleets carried treasure from the Orient, including copra (dried coconut kernels used to make oil), chinoiserie such as China silk, porcelain, gold (found in small quantities in China but not much valued there), ivory, jade, pearls and gemstones, mercury, ginger, cowrie shells, and myriad Chinese consumer goods, which gave rise to the term "China trade." Some shipments included opium. Today, I own an opium pipe acquired in China by a British mariner ancestor. It still works.

Reaching port at Acapulco, the galleons' exotic cargo was unloaded and dispersed among Spanish merchants and wealthy families. The galleons were then reloaded with silver mined in Latin and South America for the return voyage, or, if time was of the essence, sent on the next outbound fleet. According to many estimates, about one-third of the silver mined in Peru and Mexico eventually traveled the Spanish Manila galleons to Asia. In Manila, silver was delivered to Chinese buyers, or to Chinese merchants as payment for exported goods. Furthermore, much of the imported chinoiserie traveled overland across New Spain to its Atlantic ports, where it once again was loaded onto Spanish galleons, this time en route to Spain. This explains why shipwreck salvage expeditions in the Atlantic Ocean have found Asian goods aboard wrecked Spain-bound galleons. Including opium pipes.

Poor navigational skills and treacherous weather weren't the only obstacles to safely crossing the great oceans. Pirates, freebooters, and their semi-legal counterparts, privateers, lurked in the oceans, waiting to attack anything Spanish, for Spain ruled the Caribbean's bounding main and the Pacific's China Trade route, holding a monopoly on trade with the Americas. But by the end of the eighteenth century, Spain was in such heavy financial debt that frequently its fleets were escorted back to port by warships of the debtor nations. In 1778, the last *flota* sailed; soon thereafter, Spain officially opened trade in its American colonies. Soon, British and American vessels began wrecking in storms off the Pacific Coast, and even as the New World's settlers began documenting each new disaster, the legend of Spanish treasure wove itself deep into the natives' psyche.

Whose Beeswax?

In 1813, Alexander Henry, chief trader of the Northwest Fur Trading Company, writing about trade between whites and native tribes at Fort Astoria on the North Pacific, described a party of Indians offering beeswax for trade. Apiculture not being a native pastime, Henry and colleagues inquired from whence came the beeswax. The natives talked about a shipwreck off the coast south of the Columbia River, now Oregon, and described what might have been a Spanish galleon. They told stories passed down by their ancestors of beeswax floating ashore in big chunks. Soon the white settlers joined the natives beachcombing along the northern Oregon coast.

Slabs of beeswax, one measuring 24 inches by 16 inches by 4 inches, were found south of Astoria on a sandbar at the mouth of the Nehalem River. According to J. S. Diller of the U.S. Geologic Survey, who prepared a report on the beeswax dated 1895, some of the beeswax was buried in deep sand at the current high tide level. Some chunks weighed as much as seventy-five pounds, while others were smaller and shaped like candles. Diller's local guide, identified as a Mr. Edwards, reported that he had personally dug up almost three tons of the wax, some from as deep as ten feet beneath the sandy surface. (Shifting tide lines and beach erosion have buried the site of the Edwards dig beneath more sand.) By 1908 reports of beeswax finds along the Nehalem's mouth reached a total of twelve tons, including a shipment of six tons reportedly sent to Hawaii in 1847. Numbers had been carved into the larger pieces, perhaps identifying the galleon they came from, along with markings resembling religious symbols—crosses, for example. Legends of Spanish galleon beeswax spread along the Oregon coast; meanwhile, skeptics rolled their collective eyes.

Then, in 1961, a popular West Coast flotsamist, Burford Wilkerson, of Tillamook, Oregon, produced a sample of the beeswax he'd found washed up on a beach and subsequently had it carbon-14 dated. Test results put the year of origin of the beeswax at 1681, plus or minus 110 years. A chemical analysis Wilkerson had done identified the wax as being of Asian origin.

Authentication of the beeswax as Spanish galleon cargo incited a wave of beachcombing for more treasure. A list of missing Spanish Manila galleons was drawn up. Six galleons from the Manila fleet (five of them named for Spanish saints) had been lost at sea between 1571 and 1791: the *San*

Juanillo, San Antonio, San Nicholas, Santo Cristo de Burgon, San Francisco Xavier, and *Pilar.* The *San Francisco Xavier,* lost at sea in 1705, is thought to be the galleon carrying the beeswax; whether it sank at sea or wrecked on the rocky Oregon coastline before going down is still unknown. Other evidence of the galleon wreck included Chinese porcelain shards dating from the Ming Dynasty (1368–1644). But the discovery nearby of teak timbers and a ship's hull built from other exotic woods may, some experts claim, be evidence not of a galleon wreck, but of yet another Asian junk that had traversed the Pacific Ocean, whether or not its navigator had intended to sail that far.

The flotsam remains of at least sixty East Asian junks have been plucked out of the Pacific Ocean or have stranded on beaches as far north as Siberia and as far south as Southern California. Twenty-seven of the wrecked junks were discovered floating at sea, while the remainder washed up in the Aleutian Islands chain, along Alaska and the Pacific Coast, and in Hawaii. Artifacts from the wrecks have been dated from 1617 to 1876. While this evidence demonstrates that drift objects off the Japanese coast can get caught up in the clockwise circular North Pacific currents, it doesn't explain the beeswax. In the 1870s, historian S. A. Clark, while exploring the Nehalem River mouth, documented at least two shipwrecks there, but neither could be definitively identified as Spanish.

In spite of no definite proof a Manila galleon—or galleons—wrecked near the Oregon coast, Burford Wilkerson's carbon-14-dated beeswax silenced the cynics. Meanwhile, discoveries nearby of exotic driftwood, teak in particular, from ships' hulls and housings have since fueled beachcombers' fantasies, and believers in Spanish treasure are the first out on Nehalem spit following a good storm. Flotsamists know that powerful rollers can uncover tons of sand in one good Pacific storm, setting long-buried artifacts afloat.

Loaded with treasures from the Orient, or from the plundered mines of Central and South America, the sunken Spanish galleons on both American coastlines provide tantalizing fodder for individual beachcombers, and, too, for organizers of million-dollar treasure hunting expeditions. The Internet offers dozens of Web sites of diving pros raising funds to salvage shipwrecked vessels, and most concentrate on the Atlantic Coast, where divers less civic minded than Delaware's Dale Clifton trace a piece of lagan to its lucrative underwater source for the sole purpose of striking it rich.

Pirates, Pillagers, and Mooncussers

Pirate legends abound in oceanside communities. A famous tale from Oregon tells of William Charles Morgan, who from 1645 to 1657 worked the piracy route along the North American Pacific coastline in his schooner the *Inferno*. One-Eyed Willie, as he was known from Baja to Canada, got his nickname by losing an eye while battling the Spanish Navy.

One-Eyed Willie had a Mexican lover whom he deserted when she became pregnant, reportedly with his child. Willie wasn't only insensitive; he also had a mean streak: Legend claims Willie was the brains behind the invention of execution by plank walking. So meticulous was he about the *Inferno*'s upkeep, Willie worried that tossing prisoners overboard would damage his ship's hull, so he rigged the gangplank over the side, its end hanging far out over the water, and ordered his prisoners to walk to the end and keep walking. He should have patented the gimmick because it made quite a splash in the pirate trade.

British Naval officers wrote in their diaries of confrontations with One-Eyed Willie, including claims the Brits sunk the *Inferno* at sea. However, other documents record that the Brits drove the *Inferno* into a bay on the Oregon coast, at what is now Ecola State Park; firing explosives, the Brits collapsed the cliffs around the *Inferno*, but Willie and his crew allegedly survived, living in cliffside caves, where they buried their treasure.

In 1935 a gentleman by the name of Chester Cobblepot publicly claimed to have located One-Eyed Willie's treasure; Cobblepot subsequently vanished and rumor had it he had eloped with the treasure and was living grandly in some exotic locale. But in 1985, youths exploring a series of coastal caves found Cobblepot's remains. Cobblepot apparently had been the victim of a cave-in. The youths also claimed they found the *Inferno* intact in an underground cavern. Another great flotsam story, but was it true or merely the stuff of teenage fantasy?

News reports from the time speculate that the reason the *Inferno* couldn't be located by later investigators was that the tide swept into the cavern and pulled the *Inferno* back out to sea. Then in 1998 at Loomis State Park in Washington State, a mass of flotsam comprised of splintered wood, human bones, and tarnished brass washed up along a hundred-foot length of tide line. Some claimed the debris came from atop a cliff, from a beach shanty, but the shanty site was never found. Others insist it's the wreckage of the *Inferno*.

Shortly after the flotsam washed up, coastal residents started reporting sightings offshore of the *Inferno* and its crew, ghostly apparitions of starving men clinging to a three-masted schooner sailing at full mast, usually at night and usually enveloped in the typical gray haze along the Pacific Northwest Coast.

One-Eyed Willie's hidden treasure has never been recovered, but flotsamists frequently visit Ecola State Park, many of them armed with metal detectors.

According to Cape Cod legend, in 1715 a young English sailor named Samuel Bellamy persuaded a wealthy New Englander to finance a salvage operation off Florida's coast. Evidence of ill-fated Spanish galleons had washed up on Florida beaches, and Bellamy wanted to follow the flotsam to sunken treasure. Before leaving Cape Cod on his adventure, Bellamy bragged to Maria "Goody" Hallet, one of the Cape's great beauties, that he would return eventually as captain of the world's tallest and longest sailing ship. Failing to locate sunken treasure, and loath to return to Cape Cod in shame, Bellamy turned to piracy, soon earning the moniker Black Bellamy. In less than a year, Black and his formidable crew of pirates had waylaid and plundered over fifty sailing ships.

One morning off Cuba's coast, Black and his crew seized a three-masted galley, the *Whydah*, complete with a shipment of thousands of gold and silver coins, ivory, and indigo. At one hundred feet in length, the *Whydah* became Black's flagship, and he set sail for Cape Cod to show Goody Hallet what he'd brought home. But the *Whydah* was a top-heavy vessel, susceptible to high winds. In April of 1717,

Black's ship encountered a storm with seventy-mile-an-hour gale force winds and forty-foot waves. The *Whydah* capsized and broke in two.

Black went down with the ship, but two of his crew survived—an Indian pilot who is lost to history and a Welsh carpenter named Thomas Davis, who, unsurprisingly, gave full account of Black Bellamy's adventures and ultimate demise. Black lives on in the hearts and folklore of Cape Cod, and in 1984, deep sea explorer Barry Clifford discovered the *Whydah,* the first authenticated pirate shipwreck off the North American coastline.

Flotsam, jetsam, and lagan offer vital clues to the sea's secrets. Sometimes a piece of lagan leads to the solution to an ancient maritime mystery. Take the mystery of the Spanish galleon *San Martin,* swamped and sunk during a violent storm in 1618. The *San Martin*'s remains lie somewhere off the Florida coast. But where? After centuries of gold coins washing up along Florida's Wabasso Beach, the *San Martin*'s hull was discovered in the 1960s when two young lobster fishermen found four cannon submerged in shallow water. When raised, one cannon revealed its date of manufacture as 1594. Beneath where the cannon had lain submerged was the hull of the *San Martin,* five hundred to fifteen hundred feet offshore of Wabasso Beach. Professional salvagers brought up more of the *San Martin.*

Then in 1993, Kane Fisher, working the site of the shipwrecked galleon, hit the flotsamist's jackpot. Fisher found the ship's astrolabe, dated 1593. Astrolabes measure latitude by sighting the angle of the horizon from the sun at its azimuth. Before the sextant was invented in the 1730s, and the chronometer in the 1760s, a mariner's chief navigational device was the astrolabe.

The wreck of the *San Martin* today lies about a thousand feet to the south of Disney's Vero Beach Resort in relatively shallow seas. When beachcombing the area, flotsamists should remember that all that glitters is not gold, and the coin you may snap up off the strand is as likely to bear the image of Scrooge McDuck as that of a Spanish king.

Even their name is terrifying: mooncussers. They pillaged and plundered ships after luring them into wrecking by placing phony guide lights on shore. Mistaking the shore lights as benevolent guides, many a ship's captain went astray, crashing his vessel on rocky shoals or headlong onto the beach. The mooncussers then swarmed over the wreckage, pillaging its hold and robbing its corpses of their jewelry.

In 1835 a violent storm off New Jersey's coastline caused a large number of shipwrecks. In the shore town of Barnegut, rumors spread that some locals had been mooncussing. In Trenton a grand jury was formed to address the matter; it brought indictments against forty people. Among the mooncussers were two justices of the peace who were charged with plundering cargo and materials off the *James Fisher* and the *Henry Franklin*, two ships that wrecked off Barnegut.

On February 28, 1835, the Key West, Florida, daily *Inquirer*, savoring an opportunity to admonish stuffy northerners who had lately been warning of Florida's mooncussers, wrote in a scathing editorial: "We have seen numerous accounts of the shocking depravity of the persons engaged in wrecking on the New Jersey shore, and we believe that those upon Long Island are not much better. Cargoes are pillaged, passengers and others robbed, and an utter disregard of all the moral qualities of our nature manifested, and that, too, in a section of the country which is thought to be particularly enlightened, and where the laws are looked upon with peculiar reverence. Truly gratifying is it to have in our power a comparison between the acts of those engaged in . . . responsible business on the coast of Florida, and the acts of these wreckers of New Jersey and New York."

In 1846 another storm wrecked several ships along the New Jersey coast. Newspapers ran grisly stories of those who perished and blamed the shipwrecks on the same gang of Barnegut mooncussers. The newspapers reported that the mooncussers ignored the injured and drowning passengers, turned their backs on castaways, and later demanded money as ransom for delivering to families the corpses of their loved ones. Several news articles accused Barnegut residents of shining lights to mimic the Barnegut lighthouse and confuse the ships' navigators. The governor of New Jersey called for an investigation. The conclusion: No evidence proved mooncussing had occurred.

All along the Eastern Seaboard, residents of seaside villages had only limited equipment and resources for saving shipwrecked passengers. Many legends of great heroism by shore residents have been passed down in coastal lore. On the other hand, stories abounded of bodies washing up and being stripped of clothing, jewelry, and other valuables. Soon, nickel novels and cheap magazines began publishing mooncusser stories, including a novel by Charles E. Averill, *Ship Plunderers of Barnegut*, in which Averill called the village the "scene of the dread deeds of death and depredation upon the unfortunate castaways of the sea." Averill's book was a best-seller.

In 1939 a New Jersey physician, Dr. Newell, invented the "lifeline" device for saving victims of shipwrecks. Newell's lifeline was fired from a shoreside cannon to the ship in distress; passengers lucky enough to grab and hang on were then dragged safely ashore. New Jersey, especially humiliated by the Barnegut mooncussers, became known for its efforts in saving victims of maritime disasters.

The Victorians

The reader may recall an earlier reference to my night in Leonard Bernstein's bed, wherein I commented that the maestro's bunk ranks second in my personal World's Best Beds ratings, first place awarded to the moss bed I often enjoy in Rome. The Rome bed is inside a palace neither near to nor far from the Pantheon. A mattress made of Irish moss flotsam has no business popping up in a palace, even if it is Rome. But there it is. The popular travel guide and television personality Rick Steves once managed to tease from my tight lips—tight on scotch—the palace's name, but he failed to extract the exact whereabouts of the special bed. An Irish moss bed is a thing of beauty, meant to share only with one's most intimate friends. Steves, a veritable treasure trove of European history, is a likable guy, but he's always outing these precious secret places in his guidebooks. I won't have ten million tourists sharing my Irish moss bed.

Sea moss flotsam is a far superior stuffing than feather down, which holds moisture and often assumes a musty, mildew odor. Sea moss, once dried, does not attract moisture and so remains mildew free and seems to survive what beds must through the ages, with rather more resiliency than down. I do not know the age of the Irish moss bed in the Roman palace, though I suspect it dates back to Victorian times, which is where all this moss flotsam is leading us.

In the 1890s, in summer, the Victorian woman's preferred flotsam was sea moss. As it floated ashore, ladies in long skirts and wide aprons quickly gathered it up from the tide line and placed it in their aprons or gunny sacks to carry home, where it was laid out in the sun to dry. Once the green and brown mosses had dried, the ladies stuffed it inside silk pillow casings, which they then embroidered, often with clever images of seashells or nostalgic beach scenes. Crafted at summer's end at the ladies' seaside homes, the pillows traveled along with their makers to their town homes, where the mosses' fresh sea scent reminded a reclining lady of her summer idyll, no doubt bucking up her spirits during those nasty urban winters.

Time dragged in winter, and Victorian ladies of leisure needed something to do while their servants made up the moss beds. These upper-class damsels took up the art of moss pressing, flattening strands of sea moss between the pages of heavy books. Once dried, the pressed moss, resembling sea-scented lace, was pasted onto the ladies' personal stationery, providing a fragrant frame for their most intimate correspondence. Idle fingers, after all, are the devil's digits.

Which reminds me of my Scots great-grandmother. During the nineteenth century, sailors on long voyages whiled away boredom by creating valentines made from seashells, which, upon their return home, they presented to their true loves. One clever sailor figured out that by slicing a seashell it could be glued in flat cross sections into a shallow box, its crosscut design even more compelling than the uncut shell. Soon thereafter, the purple cowrie became the sailor's valentine shell of choice; slicing reveals its sexy violet interior, a phallic column seeming to penetrate swollen virginal space, or for the more puritanical, a crosscut of male and female organs united in the procreative position. Very suggestive, very Victorian. And my long-in-the-tooth great-grandmother Nell had one of these sailor's valentines on her bedside table.

The White Bullock of Bombay

The following flotsam tale fascinated me the first time I read it, the more so because had things gone differently, I would not exist. Captain Alfred J. Green, Master Mariner and skipper of numerous British sailing ships—also my great-grandfather—writes in his ship log of a murder mystery worthy of Sherlock Holmes, involving a floating corpse:

"One lazy Sunday morning in 1870, while second mate, I was being rowed ashore from the ship which was lying well offshore in Bombay harbor. I was reading a book, glancing up now and then to steer clear of the harbor traffic for I was the tiller in the stern sheets. One of these intermittent surveys of the harbor revealed something white, a point or two off the bow, and I eased the tiller a bit to bring us closer to it. A white bullock, thought I, one of hundreds of such beasts that each year are dumped into the harbor. But as we drew closer, I saw that it was not a white bullock, but a human body wrapped in a counterpane with feet and neck tied securely with sennit. A weight had been attached but the sennit had given way and the body had floated to the surface trailing the open noose that had been attached to the weight. A long tear in the shroud gave a glimpse of an officer's sleeve, showing that the murdered man was a second mate. At my command, the apprentice in the bow had taken up the boat hook to fend off this loathsome thing, but being more enthusiastic than cautious, he plunged the boathook directly into the floating mass and with an outpouring of almost overwhelming gases the body slowly sank out of sight. The case was reported to the authorities ashore, who stated that a second officer from one of the ships had been reported missing some two weeks before, and that, undoubtedly, it was his body we had sent back to the bottom of the harbor.

"The entire incident was forgotten during the next seven hectic weeks of docking, unloading, awaiting a cargo, loading and finally rounding up a crew and sailing. We were bound for Hong Kong, and after an eventual passage down the Indian West Coast, we crossed the gusty Gulf of Monar, and rounding Ceylon started our long trek across the Indian Ocean. Dondra Head was barely out of sight when the first intimation of trouble came. I had left the poop at two bells in the midwatch to go forward and find out why the fo'c'sle head lookout had not answered the bells. No cry of 'all's well' had drifted aft, and remembering that the lookout had been a big six foot four brute of the surly, troublemaking type, I was filled, to say the least, with some apprehension and an almost certain knowledge of what I would find. I was dead right. There on the fo'c'sle head, his head resting on a coil of rope and a tarpaulin pulled up to his neck for a blanket, was our lookout snoring stentoriously. I quickly went aft to my cabin, where I procured my revolver, and calling two of the watch who were in the waist and telling them each to draw a bucket of water from the harness cask, we made our way to where our friend lay sleeping. I gave my orders. As I tore off the tarpaulin, two buckets of stinking, salt-pork tainted water were doused full into the face of the giant. With a bellow of rage, the sputtering, gasping sailor jumped to his feet to be met by the glint of a leveled revolver which, while it held back his body, did not put a leash upon the volley of profanity that came spewing out of the man's very soul: 'You damned little—you—I'll kill you if it's the last thing I do, and you won't be the first whippersnapper of a second mate that has felt my hands on his throat.' Something clicked in my mind. 'That's right,' I said, desperately trying to speak quietly, 'and the last wasn't many weeks ago, was it—when you wrapped your victim in a white counterpane and tied his face and neck with sennit! Not cord or spun yarn, but sennit! Sennit that didn't hold for very long the weight you had tied to the body before you threw him into Bombay Harbor.'

"It was a shot in the dark but the man's next words proved to the three of us listening that the shaft had driven home: 'How the hell did you know—who told you—you—.'

"Second Mate Green ordered the sailor below and went aft to lay the story before the ship's captain. The skipper, a genial, fatherly old man, always saw the best in everyone, and inwardly cringed from violence of any description.

"'Now, Mr. Green,' he commenced, 'we mustn't be hasty in this—the fellow is probably scared stiff, and we'll report him to the authorities in Hong Kong. We're shorthanded and need every man aboard, so let's forget it for awhile.'

"Three or four weeks passed with the sailor proving quiet and obedient. Then, in the Straits of Malacca, less than a mile from Sumatra's jungle-clad cliffs, the sailor made his move. Second Mate Green was standing on the poop deck by the wheel, scanning the shoreline through binoculars, when something tore the cap from my head, striking my shoulders at the same instant, and I spun around to see a marlin spike quivering in the oak planking at my feet. Four inches—even two inches—closer and I would have been instantly killed. I looked aloft to see the giant of a man on the yard directly overhead, while at the same instant, the captain, who had witnessed the whole thing, roared out to the mate just appearing on deck to, 'Clap that man in irons! The scoundrel! Clap him in irons, I say!'

"But the killer sailor had no intention of spending the rest of the cruise in irons. As they neared the Sumatran shore, the sailor managed to jump, feet first, overboard. He surfaced, and struck out for the beach. While the skipper and a mate rushed to the cabins to get their guns, the others watched in horror as a dorsal fin appeared, fast approaching the swimming man. A second later, with a scream heard even by the Captain in his cabin, the shark struck, and the swimmer was jerked beneath the surface. He never reappeared."

IV. Flotsam Science

According to Norse mythology, a salt mill is grinding away somewhere at the bottom of the sea, ... which explains why the sea is salty. The discovery in the 1970s of hydrothermal vents that emit vast amounts of minerals into the sea suggests we would do well to re-examine our myths and legends.

Dr. C.
Oceansonline

Flotsam Density

Only a true flotsamist appreciates the fascinating floating qualities of what washes up on ocean beaches. The laws of density sound simple: What is heavier than the water it displaces will sink, and what is lighter than the water it displaces will float. But as the seahorse that floats somewhere between the water's surface and the seafloor so aptly demonstrates, variables matter. And density is everything.

Lava lamps taught a generation of Americans about density. Seawater has a density of 1.03 grams per milliliter. What sinks in freshwater with its 1.00 gram per milliliter density may float on seawater. Ice, with a density of 0.92 grams per milliliter, floats on both freshwater and seawater. Wood lighter than the amount of seawater it displaces will float, sometimes for thousands of miles without sinking. Objects placed into seawater will find their own level depending on their densities and volumes. The old Ivory Soap maxim, "It floats," was explained by the product's "99.44 percent pure" quality. The remaining 0.56 percent is air whipped into the soap ingredients to make the soap bar less dense than freshwater.

Spheres float, up to a point, depending on the volume of material in relation to diameter. An object denser than the water it's placed into will sink, either suspending in the water or, if dense enough, sinking to the bottom.

Flotsam can float temporarily, as in the case of a human corpse on seawater before its lungs fill with water and its density gradually increases until it finally sinks. But, as the White Bullock of Bombay illustrated, as the decomposing body fills with gases, it rises to float once again on the surface. Or flotsam can float indefinitely, as in the case of any object that is both lighter than seawater and impermeable (such as one made of plastic) or whose density even when suffused with seawater is still less dense than the water it floats in. Driftwood, displacing relatively small volumes of water, will float, as will a ship's hull until it capsizes and fills with seawater, thus reclaiming nearly all the volume of water it had displaced. Then the ship sinks.

This reminds me of a recent mocking held on my behalf by four men of the sea who shall remain anonymous. I happened to mention that I'd found this deeply rusted wrought-iron terrace table washed up on the beach and they burst into pealing laughter. I recall the most seafaring of the three actually teared up. "What do *you* know?" he taunted me.

"That table was definitely tossed over the bulkhead by some guy living above the beach." "But it had seaweed on it," I said. "And the rust . . ." He wouldn't let me finish. The four mariners laughed and said I had a lot to learn about flotsam and jetsam. I said, "What about lagan? Can we talk about lagan?" But they weren't listening. A more important subject had arisen, baseball, as I recall. We were crammed into a car at the time, and I was crushed between two of these gorgeous hunks in the backseat—can't complain about that part. I felt my face turn hot, and though I didn't pursue the lagan angle with them, I later returned to the beach to fetch the evidence, but it was gone—apparently the fellow living on the bulkhead had retrieved it.

A wrought-iron coffee table, because of the density of the iron frame, will sink. So then, why did I find one washed up on the beach? Certainly, it might have been chucked over the bulkhead from land. But the iron was, as I said, deeply rusted and seaweed-coated. Furthermore, I had walked this beach the previous day and the table wasn't on the tide line. When I discovered the table, it lay exactly along the high tide line.

Powerful undercurrents can shove heavy objects upward, and for a time they become flotsam, then sink again, often tumbling along the ocean floor until they make landfall. Once again: Jetsam can become lagan and then become flotsam, then lagan again, until a forceful current lifts it into a cresting wave headed for shore. I feel better now.

Currents, Whorls, and Gyres

Flotsamists around the world know that ocean currents are at the bottom of what the tide brings in. At the risk of spoiling the magic and romance of beachcombing—though it hasn't for me or any other devoted flotsamist—knowledge of the oceans and familiarity with ocean currents has helped serious flotsam collectors locate the best stuff that incoming tides toss up onto beaches. Those who prefer clinging to romantic legends or mystical theories about serendipitous appearances of bottles with messages inside, or prophetic messages nature carved just for you into a certain length of driftwood, may prefer to skip this section altogether. But if you can abide statistics and hang onto your romantic beachcomber genes, herewith the basic science behind the flotsam, and if you make it to the very end, you'll be rewarded with a true flotsamist's approach to beachcombing.

As we have seen, botanists and biologists have for centuries believed certain plants and animals migrated across entire oceans clinging to drifting tree limbs. This may be true in the case of Fiji's crested iguana, which apparently washed up on a South American beach eons ago. The potato, according to some accounts, traveled on driftwood from Peru to Polynesia even before the Spanish conquistadors brought it by ship to Europe. These accidental voyages have long fascinated oceanographers, especially those studying ocean currents. For decades, oceanographers have thrown bottles with notes inside into the sea, the notes asking the finder to reply. Drift bottles proved a boon to oceanographers, helping them develop mathematical models of winds, currents, and weather patterns. But drift bottles, though still used today, were eventually improved upon; floating cards—drift cards—proved more practical, as did the concept of tracking ships' cargo accidentally jettisoned into the oceans during storms. When computer technology entered the world of oceanography, it revolutionized the study of ocean currents, and the man responsible for changing the way we think about the oceans and their flotsam was oceanographer W. James Ingraham Jr., who developed the Ocean Surface Currents Simulation (OSCURS) computer model.

Jim Ingraham borrowed a computer model previously applied only to fishing and created OSCURS. Ingraham had long been fascinated with the science of tracking ocean currents, vital to the fishing industry, to overseas transport, and to sailing—even to aircraft flying over some parts of oceans during climate shifts. Because ocean currents can vary greatly from one year to the next, drift bottles and drift cards and where they are retrieved can explain unusual activity in currents. Like people, ocean currents shift position now and then. A shift in an ocean current can make a critical difference in predicting weather patterns such as El Niño and La Niña and in locating anything that moves partly at the whim of the currents, such as derelict vessels, lost sailors, marine mammals, schools of fish, downed weather balloons.

Ingraham's OSCURS technology is able to simulate wind-driven currents rippling across the ocean's surface. Using mathematical equations combining the influence of large-scale, relatively stable currents with the influence of surface currents measured in daily wind variations, and at times using the science of hindsighting, OSCURS can predict movement in the upper layers of the ocean surface, whether it's moving water or a moving target like a drift bottle or a little pink plastic propeller.

Teaming up with his University of Washington colleague Curtis C. Ebbesmeyer, Ingraham set about tracking flotsam as a way of understanding ocean currents, including the giant circular currents known as gyres. Since no single organization routinely monitored shifting currents, Ingraham and Ebbesmeyer soon established a reputation as experts in the field. The formula may seem simple: winds + flotsam + point of departure + point of retrieval, factored by the direction of average ocean currents = present path of current. They had learned from years of experience that as the distance of the release site from the shore increases, the number of bottles (or whatever was jettisoned into the ocean) recovered onshore decreases. For example, when drift bottles are released within a few miles from shore, about half of them will be recovered. If the release site is hundreds of miles out, the recovery rate can drop to as low as 10 percent. Drift-bottle studies have released as many as 150,000 bottles, although most drift studies involve much smaller numbers. Simple enough to do, but costly, and tedious.

Meanwhile, cargo vessels across the globe caught in violent storms were accidentally jettisoning tons of floating objects into the oceans. One day, according to Ingraham, Curt Ebbesmeyer's mother, Genevieve, was reading the newspaper and saw an article about flotsam washing ashore. She said to Curt, "You should study this stuff."

"That's what I do," Curt replied.

"I'm not talking about what you put into the ocean," Genevieve said. "I mean, why don't you study the flotsam that's already there?"

From that day, according to Ingraham, the two men began tracking ocean flotsam around the globe, including ships' cargo lost during violent storms. Both oceanographers are now retired, although Ebbesmeyer publishes the quarterly newsletter *Beachcombers' Alert!* out of Seattle and has subscribers around the globe who report on what they've found washed up. A disciple of Tadashi Ishii, the Japanese flotsam expert who wrote the definitive *Encyclopedia of Flotsam*. Ebbesmeyer is less eccentric than Ishii and tends to focus on the more scientific aspects of flotsam, contributing valuable data to the field of oceanography. At the same time, his folksy newsletter reaches out to even the most amateur beachcomber.

The Great Garbage Patch

Marine environmental researcher Charles Moore might begin his explanation of ocean gyres by comparing them to toilets in which the water constantly whirls but never flushes. Or, he might use the more tactful "gentle maelstrom." He should know. He has spent more time than any living human inside the gyres, researching them and all they contain.

Captain Moore was sailing home to Southern California from Hawaii after taking third place in the 1997 Transpacific yacht race from Los Angeles to Hawaii. At the helm of *Alguita*, his cutter-rigged ocean research vessel, Moore usually sought to avoid the North Pacific subtropical gyre, an area of ocean approximately the size of Africa, about 10 million square miles slowly swirling beneath an undulating high-pressure air system. Centered slightly north of the Transpac racecourse, halfway between Hawaii and the mainland's West Coast, the North Pacific subtropical gyre has about as much wind power over it as a baby's breath.

Savoring their Transpac showing, Moore and his crew decided to take a shortcut home, a route that would bring them straight through the wind-poor gyre most Pacific sailors avoid. *Alguita*, an aluminum-hulled catamaran, was equipped with auxiliary twin diesel engines and carried extra fuel, so Moore wasn't worried about lacking wind to propel the sails.

The North Pacific subtropical gyre is oval-shaped, measures about eighteen hundred miles from south to north, and reaches from Asia to North America. As its surface waters swirl, they make approximately one rotation every three years. Depending on the season, the tilt of Earth's axis, and weather, the gyre alternately bumps up against Asia or the North American mainland. The largest of the world's ocean gyres are the North Pacific subtropical gyre and the North Atlantic subtropical gyre, which includes the mesmerizing and mysterious Sargasso Sea.

The center of each gyre is similar to the eye of a hurricane; in a gyre, the water whirling at its center sits slightly higher than the surrounding water. Unlike a whirlpool that spirals downward, ocean gyres just keep spinning around and around. The gyres are created by masses of airflow moving from the tropics toward the polar caps. North of the Equator, ocean gyres travel clockwise; south of the Equator, they move counterclockwise. Air in the North Pacific subtropical gyre becomes heated at the equator and rises above the cooler air masses surrounding it. Meantime, Earth is rotating, moving the heated air mass westward as it is rising, then eastward once it cools. When the air descends at around 30 degrees north latitude, it becomes a huge clockwise rotating mass, which in turn causes a high-pressure system throughout the region. The circular winds produce circular ocean currents, which spiral into a center. As in a hurricane, winds near the center are calmer. Any vessel depending on wind power could get stuck in the still atmosphere and never escape—a nightmare scenario many mariners have feared over the centuries.

For days on end *Alguita* encountered no vessels, no land, nothing but the huge Pacific rolling out to the horizon in every direction. Here in the middle of some of the remotest ocean on the planet, Moore leaned out the cabin door to idly scan the water. That's when he spotted floating plastic objects. And then he spotted more. And more. During seven days crossing the gyre, Moore says that no matter what time of day he looked out, he never saw clear water, just a carpet of floating debris, bottles, bottle caps, plastic wrappers, beach balls, and fragments of plastic.

That the North Pacific subtropical gyre was littered with refuse was known to only a handful of marine biologists and oceanographers. Two years earlier, shortly after Curt Ebbesmeyer's mom jiggled his mental light switch, Curt—with Jim Ingraham and his OSCURS technology—attended an international oceanography conference in Hawaii. Ingraham presented a scholarly paper that alerted his colleagues to the massive amounts of flotsam accumulating in smaller gyres within the world's major gyres. These swirling trash heaps, Ingraham said, hold many thousands of tons of flotsam, including spilled cargo and jettisoned trash. At a loss for what to call these whirling gyres of man-made flotsam, Ingraham and Ebbesmeyer coined the term "garbage patch." The terminology caught on: Today, oceanographers studying the North Pacific's Great Garbage Patch speak of its two subgyres as the Eastern Garbage Patch and Western Garbage Patch. The same labels are attached to similar gyres in the North Atlantic and Indian Ocean. Ingraham's OSCURS computer tracking system is able to apply hindsighting to objects floating in and out of the patches, to trace their places of origin, and to predict where they might wash up. But the extent of the garbage circulating in the North Pacific subtropical gyre was not known until two years later, when Captain Moore and his *Alguita* crew returned home from the 1997 Transpac and reported their findings.

What Moore already knew was that within the North Pacific gyre are two subgyres, each roughly the size of Texas. The western subgyre lies off Asia; the eastern subgyre lies off the U.S. mainland. While the entire North Pacific gyre was found to contain vast amounts of garbage of human manufacture, the two smaller gyres each contained thick, soupy concentrations of the same trash in a partially degrated state.

A piece of flotsam might circle in the gyre or subgyres for hundreds of years before spinning off into a feeder current. For example, take an inflated beach ball jettisoned off a pleasure boat in the Hawaiian Islands. Depending on wind direction, the beach ball may enter the Kuroshio Current off Japan, travel north on the Kuroshio and enter the North Pacific Drift, and from there enter the California Current off Washington State, then travel south on the California Current and eventually wash up on a U.S. West Coast beach. Or, the beach ball may veer off into another branch of the Kuroshio, travel back toward Hawaii, or even the Philippines. More fantastic, the beach ball may just continue drifting from one current into another without washing up, perhaps eventually reentering the North Pacific subtropical gyre for another spin around the ocean's six-thousand-mile circle of coastal currents. Once in the Great Garbage Patch, the beach ball might get sucked into the more concentrated western or eastern garbage patches, adding to what Moore calls a viscous "synthetic broth."

Infinite possibilities exist for the beach ball and other flotsam riding the ocean currents. One thing's for certain, says Moore: Made of plastic, the beach ball will never biodegrade, and the longer it drifts in the ocean, the more poisonous it becomes. Plastic, a petroleum-based substance, does not biodegrade, but rather photo-degrades, breaking down gradually into smaller and smaller bits. These bits will continue to drift on the ocean, where they often are ingested by birds, fish, and sea mammals, taking in plastic's toxins. If not ingested by sea life, the plastic will continue to photo-degrade until it has broken down to the molecular level. At this point the molecules absorb any DDT or PCBs and other toxins in the water, making the molecule even more toxic. All this would be fine, perhaps, if the beach ball were the only piece of plastic circulating in the gyre. But according to Moore's calculations, the two smaller gyres inside the Great Garbage Patch each contain six times more plastic per weight than they contain zooplankton, the principal organisms that live in the nutrient-poor gyres. This translates into billions of tons of nonbiodegradable refuse.

Moore was not exactly a novice to ocean flotsam, jetsam, and lagan. He was raised in Southern California by the Pacific Ocean and has worked in seagoing trades for more than fifty years. During those decades he watched in alarm as the amount of human refuse in the ocean gradually increased, particularly plastics. But Moore became especially worried at what he saw in the gyre. This carpet of garbage whirling around in remote Pacific waters really galvanized his interest; on that single voyage through the plastic sea, Moore calculated the weight of the gyre's litter at around 3 million tons. Then and there he vowed to return to test his estimate.

The Great North Pacific Garbage Patch was about to have its measurements taken. In August 1998, Moore and a crew of four volunteers set out aboard *Alguita* from Point Conception, California. They brought various nets including a manta trawl, a fine mesh net attached to a frame shaped somewhat like a manta ray with a wide mouth and wings. Eight days out, they reached the edge of the gyre about eight hundred miles offshore and decided to test the manta trawl. They trawled for flotsam for about three and a half miles before reeling in the net. Among a small amount of sea life, they found thousands of fragments of colored plastic. Amazed, they went out for more, using a complex series of nets. In the end, they hauled aboard about one ton of debris, including numerous large objects such as an inflatable dinghy and what Moore describes as a "menacing medusa of tangled net lines and hawsers that we hung from the A-frame of our catamaran and named Polly P, for the polypropylene lines that made up its bulk."

Moore and his crew returned to the gyre in 1999 and again in 2000, 2002, and 2005, each visit verifying the statistical data gathered on previous voyages. In September 2000 they trawled a six-thousand-mile transect across the gyre. Moore reported the gyre's surface layer contained "alarming amounts of plastic products, tons of drifting nets, plastic bags, packing straps, and common household items like soap and deodorant bottles."

He has also found Japanese traffic cones, jellyfish entangled in fishing nets, sarps and transparent filter-feeding organisms with stomachs filled with plastic bottle caps, a drum of hazardous chemicals, an inflated volleyball coated in goose-necked barnacles, a cathode-ray television tube (nineteen-inch screen), an inflated truck tire mounted on its steel rim, a gallon-size bleach bottle so brittle that it crumbled in one crew member's hands. Each time they dove to visually confirm their findings at the end of a trawl, they found the ubiquitous soup of plastic fragments. On one trip into the gyre, on September 7, 2000, Moore found what appeared to be a container spill of plastic bags that covered more than ten miles of the gyre's center. The bags bore the imprints of Taco Bell, Sears, El Pollo Loco, Fred Meyer, the Baby Store, and Bristol Farms. A very large "mother bag" was also found, an indication the large bag had fallen off a cargo vessel en route from manufacture in Asia to the United States.

According to various visitors, the Great Pacific Garbage Patch also contains refrigerators, air conditioners, construction materials, chairs, glass fish-net floats, fenders, tires, wooden crates, athletic shoes, flip-flops, glass sake and Suntori bottles, even bottles with messages inside, and ghost nets that trap and kill thousands of marine species and birds.

Jim Ingraham's OSCURS drift charts show flotsam originating off Pacific Rim coastlines trapped in currents that will take the flotsam on a virtually endless odyssey around the ocean. What doesn't snick off into a side current and wash up, that is. Employing OSCURS simulation and long experience, Ingraham and Ebbesmeyer have correctly predicted when and where whole cargo containers of spilled debris will wash up. Some of the flotsam washes up in Pacific bird colonies, such as those of the Tern Island albatross, the Guadalupe Island albatross, and Laysan albatross. In the Algalita Foundation's film *The Synthetic Sea*, Bob Dieli, a Midway Atoll park ranger, shows the carcass of a Laysan albatross and points out what was found in the dead bird's stomach: a large screw-top plastic bottle cap that appeared to be from a shampoo bottle, a light stick used for fishing nets, and an electrical wire plug. Moore and his colleagues have seen in the last decade an "alarming increase" in the numbers of seabirds killed by human-generated garbage. Moore says, "Ours is the last generation of human beings who will remember what the ocean was like before it got trashed."

As for the sea life, "For jellyfish and other marine animals, it's like putting them on a plastic diet. It becomes part of their tissue."

Moore points out that once these creatures ingest the toxins, they are eaten by fish and the toxins pass into the food chain, eventually ending up in the food consumed by humans.

Rob Krebs of the American Plastics Council says, "Just because it's everywhere, it shouldn't be the whipping boy of environmentalists." Krebs correctly states that plastics do "so much, so well," and adds, "and so when we talk about Charles Moore we really have to look in the mirror." In other words, it is not plastic that pollutes, but people.

And Moore partly agrees. "It's everybody's fault. There are no single guilty parties in this problem of plastic flotsam. Until the world decides to convert most of its petroleum-based plastics to something biodegradable, we're going to live with this problem, and as more plastics are used, the problem is just going to get worse. But there are biodegradable products out there, made from corn oil and soybean oil. People have to make the conversion."

The Slick Sargasso Sea

In the film *The Wide Sargasso Sea*, a woman drowns among ribbons of seaweed in sunlight-infused waters, a scene shot from below the surface, offering a fish-eye view of drowning. But the tragedy is lost to the beautiful mystery of languid sargasso seaweed slow-dancing as it absorbs sunshine in a sultry windless gyre. Located within the North Atlantic subtropical gyre—the North Atlantic's Great Garbage Patch—the Sargasso Sea is not all drowning beauties and sargassum. Like the North Pacific Garbage Patch, the Sargasso Sea, according to Captain Charles Moore, is thick with flotsam from transoceanic cargo spills and other man-made products that rode ocean currents before entering the world's most baffling doldrums. Some visitors—and there aren't many—insist they've seen satellite parts, space shuttle debris, deflated weather balloons, airplane parts, crash debris, crates of plastic wrapped herion, and scores of Evian bottles. It's a flotsamist's dream come true, a garbage patch whose waving viridescent fingers hold fast whatever slips into their grasp.

Bounded by the powerful Gulf Stream on the west, the North Atlantic Current and Canary Current on the east, and the North Atlantic Equatorial Current on the south, the Sargasso Sea is elliptical, about two thousand miles long and seven hundred miles wide, covering approximately 2 million square miles of ocean. Located roughly between 20 and 35 degrees north latitude and between 30 and 70 degrees west longitude, the sea lies between the West Indies and the Azores. At 32 degrees, 20 minutes north latitude, Bermuda's pink beaches fringe the Sargasso's northwestern edge.

Known in legend as a lifeless realm, the extremely salty, almost motionless Sargasso Sea produces abundant sargassum, organic flotsam every Florida beachcomber knows intimately. The European eel leaves its larvae there among the sargassum; hatched eels pop off to Europe for a spell, returning seasonally to lay eggs of their own. Carthaginian Admiral Himilco noted the sea's existence sailing through the Pillars of Hercules: "Many seaweeds grow in the troughs between the waves, which slow the ships like bushes . . . . Here the beasts of the sea move slowly hither and thither, and great monsters swim languidly among the sluggishly creeping ships."

Not exactly a sailor's mecca. And, ahoy there, matey, you've just entered the Bermuda Triangle, where anything can happen, and if you survive and make it home to shore, no one will ever believe you, for the Sargasso Sea is too mysterious a realm to be believed. Like Las Vegas, what happens there stays there. When Christopher Columbus sailed into the mats of sargassum, he tried fathoming the sea and found no bottom; he got out as fast as any sailor can through the thick matted morass.

Actually, the sea does have a bottom, but it's miles down. Like the North Pacific subtropical gyre, the Sargasso Sea is calm, even as some of the world's strongest currents—the Florida, Gulf Stream, North Equatorial, Antilles, and Caribbean—rage around its perimeter. And although the sea rotates and even changes position with weather and temperature fluctuations, like the North Pacific gyre, because of its languor and thick seaweed matting, most of what goes into the Sargasso remains in this eerie swirling seaweed forest.

Legends abound of lost ships and sailors who deserted their vessels—but where did they go?—while the ships were later found deserted. During Spanish maritime supremacy, when a ship got stuck in the Sargasso Sea, the crew would toss their war horses overboard, thus the name Horse Latitudes. One story tells of a slaver with only skeletons aboard. In 1840 the London *Times* reported the ship *Rosalie* had sailed into the Sargasso Sea before being found abandoned, no signs of life. One of the more oft-told Sargasso legends is that of the *Ellen Austin*, a schooner whose crew was said to have spotted another schooner, this one deserted. The *Ellen Austin* put some of its crew aboard the derelict schooner and the two ships sailed in tandem toward port. Two days later, the crew of the *Ellen Austin* noted the other schooner sailing erratically. The *Ellen Austin* sent more crew aboard only to discover the second crew missing. In 1857 the *James B. Chester*, a barque, was found deserted in the Sargasso Sea. When boarded, the rescuers found stale food set out on the mess table and chairs kicked over.

Pirates? Conniving sirens? Alien spaceships?

More contemporary myths include the deserted *Connemara IV*, sighted in 1955 drifting 140 miles off Bermuda. Between 1969 and 1982, numerous yachts and pleasure boats were found deserted in the Sargasso's seaweed mats. In 1980 the SS *Poet*, bound for Gibraltar, disappeared in the Sargasso Sea. Although science and shipwrights did their best to explain away all superstition, the usual folklore spread. Why do some—but not all—of the ships transecting the Sargasso Sea disappear? The plot, as they say, thickens.

With its warm waters and thick seaweed mats, the Sargasso Sea is almost void of life farther up the food chain than seaweed—but its tangled mats of sargassum capture and retain whatever flotsam enters its realm. Regardless of the thousands of North Atlantic ship crossings each year, the Sargasso gives up less flotsam than the North Pacific Garbage Patch, which means that although it is smaller than the Pacific Patch, the Sargasso may contain the world's richest mass and variety of flotsam.

Captain Moore, who has visited and researched the world's three largest ocean gyres, has noted the same collections of flotsam in all of them, but he believes the Sargasso Sea is the most challenging of all.

Nike Flotsam

In May 1990, *Hansa Carrier*, a huge container vessel, encountered a severe storm in the North Pacific at about 48 degrees north latitude, 161 degrees west longitude, on its voyage from Korea to the United States. A large wave swept twenty-one shipping containers overboard. Five of the containers altogether held about eighty thousand Nike athletic shoes, including running shoes, children's shoes, and hiking boots. Some, or possibly all five, containers broke open and spilled shoes into the storm-tossed waters.

About six months later, around late November, during an incoming tide, someone along the Washington coast found a Nike shoe washed up on the beach. Then another, and another. In the early months of 1991, Nikes began washing up along Vancouver Island, and then more Nikes showed up farther north in Queen Charlotte Sound. The count rose and rose until it reached several hundred Nikes—not necessarily exact matches in color, style, or size. The West Coast flotsamists' grapevine reached down from Canada to Washington and Oregon, where shoes were also rolling in, and then finally to flotsamist Steve McLeod in Cannon Beach, Oregon.

McLeod, too, was hauling in Nike flotsam along the Oregon coast. Oregon's news media picked up the story, and soon the folks in the Queen Charlottes were talking to McLeod down in Cannon Beach. While the story made national news, the beachcombers were busy collecting the shoes. McLeod's apartment soon was packed to the gills with Nikes, which he carefully cleaned, dried, and debarnacled, even removing their laces, cleaning them, and returning them to the shoes they came from. His place resembled a shoe factory production line, all the shoes now clean and wearable. Only problem was, he couldn't make a lot of matches. What good is one Nike shoe without its mate? Unfortunately, although they were laced, matching shoes hadn't been tied together; when they fell into the sea, the pairs parted ways. Assuming other beachcombers had made similar discoveries, McLeod contacted fellow scavengers along the Pacific Coast through a remarkably swift word-of-mouth communication. The result? Dozens more Nikes in Oregon, the majority being left-footed.

Of the approximately eighty thousand Nikes released into the ocean, McLeod and other beachcombers collected about thirteen hundred shoes from the waves. But how to identify them as from the *Hansa* spill? And then someone noticed the shoes bore serial numbers. Smart, Nike.

At first declining to discuss the shoe flotsam, the chagrined Nike Corporation, after considerable wrangling and diplomatic exchanges, admitted that, indeed, their athletic shoes had accidentally fallen off the *Hansa Carrier* into the North Pacific, and verified this with the serial numbers. These were no black market items; these were genuine Nikes. Owners of the *Hansa Carrier* suffered similar embarrassment as maritime insurance companies investigated the accident to determine liability. Had the carrier secured the containers where they were stacked just aft of the fo'c'sle? Had the proper equipment been used to secure them?

For Ingraham and Ebbesmeyer, the Nike spill represented a great opportunity. This accidental flotsam, and the recovery of approximately thirteen hundred single shoes, dropped into their laps data they could not have planned better: The date, time, percentage of recovered shoes, and exact location of the spill would allow them to test and calibrate their ocean currents model. Using OSCURS, Ingraham and Ebbesmeyer aimed to determine the drift pattern of the shoes from the time they hit the water until they washed up on North American beaches. OSCURS suggested most of the shoes making landfall would wash up near the northern tip of Vancouver Island, central British Columbia coast, and Washington State beaches approximately 249 days after they went overboard. The first reports of shoes washing up had come from Vancouver Island and Washington beaches about 220 days following the spill. Now this is important: A

Avias galore plucked from the Oregon coast.

large number of shoes had beached on the Queen Charlotte Islands northwest of Vancouver Island. A second large number washed up farther south, at Oregon beaches. Why had some Nikes headed into northbound currents while others traveled on a southbound current? The answer: The slight toe curvature of left- and right-footed shoes caused the right-footed shoes to tack northeastward into the Alaska Current, passing the Queen Charlottes along the way, where many beached. Meanwhile, the left-footed Nikes tacked snugly into the southeast-bound California Current, and as it passed Oregon, were caught on an incoming tide.

As time passed, Steve McLeod didn't know what to do with all those Nikes taking over his apartment. He put the word out: Let's get together and match shoes, size for size, left to right. He organized a Nike shoe swap. In one day the swap paired up twelve hundred right and left shoes of the same size, left matched to right by serial number, and the pair went to whomever fit the shoe. McLeod even visited the Queen Charlottes and found more shoe matches there.

Postscript: In the summer of 1992, more Nikes washed up, this time at the northern tip of the island of Hawaii. The theory is that these shoes made a nearly complete round trip on the circling currents. In 1999 another spill tossed fifty thousand pairs of Nike cross-trainers into the ocean. OSCURS correctly predicted their washing up along the North American West Coast in 2003.

An athletic shoe can stay afloat for about ten years and is still wearable after three years in the ocean. Nike flotsam shoe matching is a continuing sport along North America's West Coast.

Post-postscript: On January 9, 2003, the West Coast's most prolific flotsamist, John Anderson, found a left-footed, blue and white Nike EZW men's basketball shoe on the beach at Queets, Washington. A week later Anderson found an identical shoe, also at Queets. The only difference between the two Nikes: One was a size 8½ and the other a size 10½. Anderson noted that the shoes were made in Indonesia and he noted the serial numbers. He consulted with Ebbesmeyer, who deciphered the shoes' codes. According to the serial numbers, Nike had placed its order for the athletic shoes on October 10, 2002, and the order was filled in Indonesia on December 12, 2002. Nike confirmed that during a storm on December 15, 2002, three containers of their shoes were accidentally jettisoned overboard forty miles off Cape Mendocino, California, while the vessel caught heavy seas en route to port in Tacoma, Washington. (A total of ten containers had fallen overboard.)

But what of the remaining seven containers that fell into the sea? John Anderson may have the answer, for along with the Nikes on the Queets beach, Anderson found some metal cans, still sealed and light enough to float.

"Turns out they were chow mein noodles," John told me. "And they were fresh. Sure, I ate some. They were good, too."

Good thing, too, because the Nikes I saw from that spill were hella-bad ugly.

Not Only Nikes

Each year about 100 million cargo containers cross Earth's oceans. Manufacturers of everything from Barbie dolls to BMWs hire shippers to navigate ocean waters between their factories and their retailers. Container vessels carrying an average of forty-five hundred shipping containers ply the Pacific every day, each container being a size that fits snugly on one railroad car or one semi-truck. Most of these cargoes reach port safely, but as any importer can attest, disasters happen. About one thousand cargo vessels each year accidentally dump some or all of their cargo into the sea, for an average of approximately ten thousand cargo containers per year falling into the oceans.

I met a woman in a movie queue who had just purchased a BMW from a European factory. Just the day before, she said, a BMW representative telephoned to say that her custom-outfitted BMW had fallen overboard into the Atlantic Ocean. "It was a nice car," she told me, "really, really nice."

"So what happens now," I asked.

"Oh, they'll replace it. The guy said, with all the stuff, just as I'd ordered it." She shrugged good-naturedly. "Hey, things happen."

Cargo container spills cause hazards at sea. On January 11, 2000, the British FV *Solway Harvester*, a scallop trawler, went down in the Irish Sea, claiming the lives of seven crew members. Plastic vats filled with mayonnaise were found near where the trawler sank, according to the Scottish Scallop Association. Investigators strongly suspect the trawler collided with a spilled cargo container.

An unfortunate importer in 1993 lost seventeen thousand hockey gloves, shin guards, and chest protectors to the Pacific Ocean in a cargo jettison. The gloves floated, traveling faster when the glove's fingers pointed up, acting as a sail.

Another importer lost a load of Legos. On February 13, 1997, as the *Tokio Express* sailed off Land's End, England, a rogue wave tossed the cargo vessel, jettisoning sixty-two containers overboard, including a container full of Legos from Denmark. That one container carried 4,756,940 miniature plastic toys, in a hundred different shapes, many of them kits for making boats. The Legos, originally bound for Connecticut, included, among other items, 28,700 yellow life rafts, 418,000 red or blue divers' flippers, 97,500 gray scuba tanks, 132,000 gray and yellow diver legs, 4,200 black octopuses, 33,941 black and green dragons, and 54,400 pieces of green sea grass.

Curt Ebbesmeyer, by now tracking cargo spills with all the glee of a mischievous schoolboy, performed a "sophisticated test" involving a bucket of saltwater, concluding that fifty-three types of the Legos floated, while the remaining forty-seven types probably sank. By Ebbesmeyer's calculation, as many as 3,178,407 Lego pieces could have gone adrift. That is, if the container actually broke apart at sea. That some Legos escaped the container was confirmed when some Lego dragons and Lego sea grass washed up in Cornwall, England. It takes about fourteen months to travel around the Atlantic gyre, the clockwise loop that the joining currents form from Cornwall down to Spain, then across to Florida into the strong Gulf Stream and up along the Carolinas. The Danish flotsam may already be washing up along North America's eastern coastline. And because Lego is such a well-made product, its durability means the little life rafts and dragons, etc., could stay afloat for decades. OSCURS predicts Legos will wash up on Alaska beaches in 2012, on Washington State beaches in 2020.

The Rubber Duck Lowdown

In 1992 a factory in China shipped out 28,800 rubber beavers, turtles, frogs, and ducks, destined for North America for a company called The First Years, in Avon, Massachusetts, where they would be sold as Floatees and placed in the bathtubs of girls and boys and even grown-ups of the nostalgic sort. The rubber bathtub toys were stuffed into plastic and cardboard packaging, each containing a yellow duck, a green frog, a blue rubber turtle, and a red beaver. The plastic was glued to cardboard backing and placed into a carton with thousands of others just like it, and the carton was loaded into a cargo container. The container was loaded onto a ship.

As the ship approached the international date line, it encountered a violent storm. The ship tossed and rolled. Everything inside the cargo containers shifted. And then containers broke from their bindings and fell into the storm-tossed sea. When the container with the toys hit the choppy waves, it broke apart, spilling the packaged playthings into the sea. Within a day or so, the glue holding the cardboard backing to the plastic housing dissolved, releasing the four rubber toys from each package into the ocean.

The first reports of rubber toy sightings came late that year. On November 16, 1992, at least six of the rubber toys washed ashore just south of Sitka, Alaska. A few days later, twenty more toys were plucked off a beach just north of the first beachings. Then, between November 1992 and August 1993, about four hundred more of the bathtub toys washed up on beaches between Cordova, Alaska, and the southeastern Gulf of Alaska beaches of Coronation Island.

OSCURS got on the case. The computer program simulated a drift model of the toys. Consulting the cargo vessel's logbook, oceanographers Ingraham and Ebbesmeyer pinpointed the release time and location. The U.S. Navy's Numerical Oceanography Center provided wind speed and direction information. Using a satellite tracking system, they constructed a simulation model, which ultimately suggested that over the next two years, many of the toys most likely would drift westward along the Alaska coastline to eventually enter the Bering Sea. From there, the only way out would take the floating rubber toys past icebergs to the northern coast of Greenland, and from there, into the Atlantic Ocean.

In 1995, ducks and turtles were sighted off the Washington coast, evidence that some of the bathtub toys traveled south. In 2003, a duck was found on the coast of Maine and a frog turned up in Scotland.

As much as insurance companies mourn the accidental jettisoning of cargo into the oceans, Jim Ingraham, forever the optimist, calls them "spills of opportunity." Both Ingraham and Ebbesmeyer view the cargo spills as boons in their effort to improve mathematical models of surface current drifts, especially those affected by seasonal variations or weather anomalies like El Niño and La Niña.

Year of the Flotsam

They come around once every ten years or so, the latest being spring 2003. They are big flotsam events—massive amounts of stuff washing ashore. Curt Ebbesmeyer called spring 2003 the "flotsam event of the century." In fact, he used the phrase as a banner headline for his July–October

2003 *Beachcombers' Alert!* newsletter. And indeed, it was a historic season for flotsam on North America's West Coast. A persistent southwesterly wind pattern combined with unusually prolific spillages at sea brought untold treasures to West Coast beachcombers. That's the year I plucked a fifty-year-old life jacket out of the Strait of Juan de Fuca, and a Timex digital watch, still working, set, perhaps coincidentally, to Japan's time zone, the digital date correct, allowing for the international date line. Timex wasn't kidding when it claimed, "It takes a licking but it keeps on ticking."

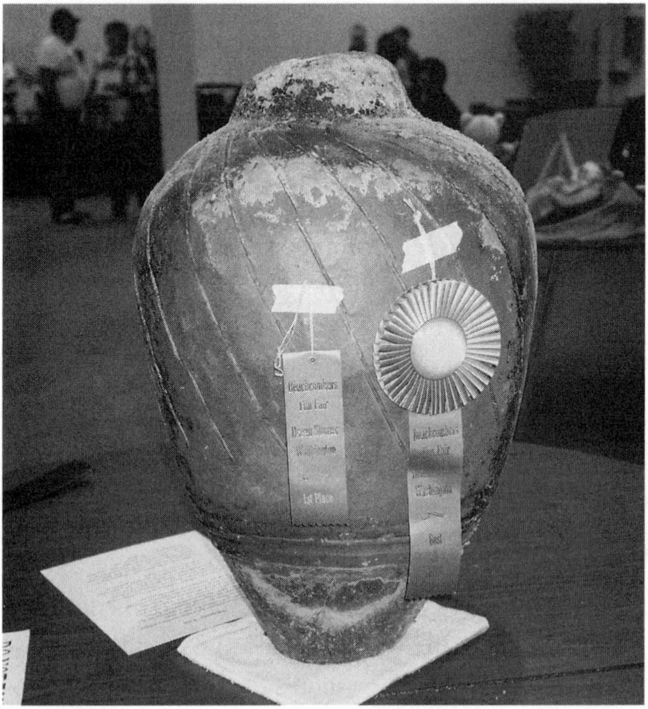

Ancient pottery vessel of mysterious origin. Blue ribbon winner at 2005 Beachcombers' Fair, Ocean Shores, Washington.

Other beachcombers simply could not haul all their finds off the beaches, so many treasures washed up. Correspondence in bottles arrived in droves, many from Japanese schoolchildren who make a hobby of tossing bottles with messages into the sea. Glass fish-net floats dating from as far back as the 1950s washed up. Whole flotillas of floatables, including ancient pottery urns, those thirty-four thousand lost hockey gloves, Tommy Pickles cartoon heads, Lightning Bolt sandals, little Legos life rafts, plastic umbrella handles, Christmas-themed items lost in the Pacific in October 1998, even bottles of soy sauce, which might have spruced up John Anderson's chow mein noodles. And Nikes galore.

Interestingly, that spring, Steve McLeod in Cannon Beach, Oregon, found twenty Lightning Bolt sandals and Brian Gisborne on Vancouver Island, in Canada, reported fifty Lightning Bolts. Left- or right-footed? Alas, no one recalls exact figures. Such is the fate of longtime flotsamists; once you understand the currents, once a lost cargo has been reported, its estimated date of landfall predicted, actually finding the cargo doesn't hold the same rush the cargo cultist enjoys. As for the rubber ducks, by now their pretty yellow bodies have faded or even disintegrated from exposure to the elements, yet some may still be out there, bobbing across the waves, and, if OSCURS's predictions hold, some now bob in the Atlantic Ocean, just off the British Isles.

As Ebbesmeyer says, "You can learn a lot from a duck on a beach."

In the summer of 2005, following a particularly ferocious North Pacific storm, I gathered an abundance of fascinating flotsam. Among the treasures were a human femur bone, a Japanese flip-flop, a barnacle-coated sake bottle—its label intact, no message or sake inside—and a barnacle-encrusted blue plastic football. I wonder about their marine adventures. Had they entered into and swirled around the North Pacific's Great Garbage Patch? How long had they swirled before catching a quirky outbound wave? If only flotsam could talk.

What exactly is gyrating around in the North Pacific, North Atlantic, and Indian Ocean subtropical gyres? Not only plastic. Whole cargo loads of detritus—that is, consumer items—are swirling around in these great whorls, just waiting for a serendipitous current to snick them off and drive them into a coastal current and then toward shore. Who knows? Another lost cargo container of chow mein noodles from China might eventually break apart, sending noodles to Nome—and everyone knows the Inuit like Chinese food—or they might veer off farther south, to wash up in the Philippines, and by then they'll need a lot of hot sauce. Or the noodles could make another journey around the ocean currents before reentering the Great Garbage Patch, there to languish well past their expiration date. The Chinese rubber frogs, ducks, beavers, and turtles that stranded on an Alaska beach certainly proved a windfall to the village's children. Flotsamists constantly wonder when the gyres will twitch off their most coveted desires.

V. Life's A Beach

The sea does not reward those who are too anxious, too greedy, or too impatient. One should lie empty, open, choiceless as a beach—waiting for a gift from the sea.

Anne Morrow Lindbergh (1906–2001)

In the early 1900s most flotsam washing up on the world's beaches was still organic in nature, and a high percentage of flotsam had been jettisoned from shipwrecks along storm-worn coastlines. In January 1900, London newspapers reported that the British steamer *Malin Head*, under the command of a Captain McKee, was sailing from Ardressan to St. John, New Brunswick, when the crew picked a piece of wood flotsam out of the ocean. It bore the letters "Merrim." The British steamer *Merrimac*, which had set out from Quebec on October 27, 1899, under the command of Captain Robert Shallis, bound for Belfast, Ireland, had disappeared after leaving Canada and was never heard from again. The piece of wood flotsam found by the *Malin Head*'s crew was believed to have come from one of the *Merrimac*'s lifeboats, and was the first sign anyone reported of the missing vessel.

On October 2, 1906, following a violent hurricane that wiped out forty-four lighthouses on the U.S. Gulf Coast, a report from Pensacola, Florida, told of the destruction of Fort McRae, its batteries and guns being swept away by the violent ocean tides. Even as "Queer flotsam and jetsam" from the sea washed inland as far as the coastline railroad tracks, five men were said to be "chained in the army hospital, raving maniacs," victims of the hurricane. Meanwhile, the coastline was dotted with fishermen's cottages, which, along with their occupants, were "as if by one stroke" washed out to sea to join other flotsam set adrift by the hurricane.

On January 21, 1910, the Dutch mail steamer *Prins Willem II* sailed from Amsterdam for West Indian ports, ultimately bound for New York. The crew and passengers numbered fifty-two. It disappeared at sea and no word of it came until March 16, when a life buoy and a boat's oar, both bearing the vessel's name, were plucked off the ocean's surface near Belle-Ile on the coast of France. Shortly afterward, the ship's propeller and a name board washed up on a nearby island.

On October 2, 1927, Captain G. A. Patterson, commanding the Nelson Line steamer *Buffalo Bridge*, noted wreckage floating in the ocean about forty-eight miles off Cuba's coastline at Guantanamo. Patterson believed the wreckage was from an airplane flown by Georgia flier Robert Redfern, who was attempting to fly from Brunswick, Georgia, to Brazil, but whose airplane had disappeared over the ocean. Besides Patterson's flotsam find, other flotsam believed to have come from Redfern's airplane was found floating in the ocean, while in northeastern Brazil, citizens reported sighting an airplane off their coastline. The flotsam offered the only evidence that Redfern's airplane had gone down over water. As forensic evidence, flotsam now included objects falling from the sky.

By the 1920s, nearly as much man-made as organic flotsam and jetsam was washing up on the world's beaches. Beachcombers began holding scavenger hunts along the tide lines, with prizes offered for the most interesting flotsam. And beachcombers also had learned that the more valuable items weren't necessarily found on a beach's surface, and that by probing a few inches into the sand, they often plucked old coins and even valuable jewelry from the beach, some of it flotsam from shipwrecks, ancient and modern. On the U.S. East Coast, a popular flotsaming beach was the Long Island strand from Nassau-by-the-Sea to Jones Inlet. After a winter storm, beachcombers bragged of collecting enough driftwood on this strandline to last the entire winter. Not only driftwood, but hewn logs of valuable white oak washed ashore here and were collected to sell at market. Finally, beachcombers became more adept at reading ocean currents, and science was about to make flotsaming a geek's pastime.

Seattle's Alki Beach, on Puget Sound, is the landing point for much of the flotsam, jetsam, and lagan washed in from the ocean via the Strait of Juan de Fuca. Scuba divers come daily to view numerous treasures hiding beneath the water's surface. Many come to visit the resident octopus; he's only about thirty feet long—small compared with other octopuses in these waters supporting the world's largest. He's somewhat shy and keeps to the colder water near a submerged natural shelf, and I don't blame him, because on any given day, a dozen or more divers seek him out. There's more to see while searching for lagan: bright violet and orange–fingered anemones; jellyfish, two feet in diameter, fringed red and ruffled; iridescent, transparent, dinner-plate-size jellies; and of course the great lion-maned medusas. There are huge Dungeness crab, salmon, six-gill sharks (one measured twelve feet long), wolf eels so tame they eat out of your hand, and sea lions, seals, otters, even migrating whales. Some local divers have lately reported counting six small sharks nearby, though these I haven't seen yet. All within Seattle's city limits.

Diving conditions at Alki are so ideal that diving schools train their students here, and police SWAT divers come here to stay in shape. Eight minutes by car from central downtown, Alki is easily accessible during an office worker's lunchtime, and so many divers come to escape the city's noise and hubbub in the silent underwater world. Yet the most fascinating dives take place at Myrtle Edwards Park and along the piers and docks lining Seattle's downtown waterfront. Here's where local history's lagan surfaces, gets plowed under, then surfaces again, until some enterprising or just plain curious scuba diver plucks the treasures and brings them to the surface. Glass bottles are a frequent find, dating from Seattle's infancy—liquor bottles, or medicine bottles, even milk bottles. Some are more than a century old; others, like the hair tonic bottle I found recently, are relatively new discards. Its cap was intact, so unlike other bottles I've found, nothing lived inside. Porcelain dishware, ceramic shards, old stoneware mugs, and dishes are among my favorite finds. Some divers go after metal—gold coins, old ship bells and lanterns, silver jewelry, anything that makes their metal detector tick, click, or jump. Every coastal city in the world is girthed with history's lagan.

Dead Men Don't Lie

War never spared an ocean. The old bones of countless sailors inhabit shipwrecks on the ocean floor, and ocean beds are littered with the lagan of naval battles fought over thousands of years. Live land mines and grenades washed up on beaches have killed and maimed thousands of children who found them. What remains of humankind's conflicts on the seas is Poseidon's secret, until the sea gives up a piece of it to remind us of our dark side.

During World War II, Great Britain, preparing to invade Sicily, needed a strategy for diverting German troops from Sicily before the invasion. Numerous scenarios were tossed around among British Intelligence officers. Then Lieutenant Commander Ewen Montagu, the point man in charge of the invasion, asked his colleagues, "Suppose I wanted to put a dead body in the sea, and let it float ashore, and have it accepted by the people who find it as the victim of an air crash at sea. What sort of body would I need?" The answer to that question sparked the idea for Operation Mincemeat, in which the Allies created the illusion that Great Britain planned to invade Greece.

In order for the ploy to work, a dead body had to be procured, phony invasion plans planted on his corpse, and his corpse set afloat at sea where the tides would wash it up on a German-occupied beach. A beachcombing German soldier might spot and retrieve the dupe. Surely dead men don't lie.

This brilliant military plan was implemented with the body of a freshly drowned young Scotsman, whose corpse was fitted out with the phony papers and conveniently jettisoned to wash up on an enemy-patrolled beach.

In 1956, this true tale of how the Brits fooled the Germans was the subject of the film *The Man Who Never Was*, the script co-written by Montagu himself. Clifton Webb starred in the role of Montagu, while Montagu himself played an uncredited cameo as an air marshal. The uncredited voice of Winston Churchill was actually that of a young Peter Sellers.

Sir Lindal's Miraculous Floating Homes

Imagine you are beachcombing when you stumble across a huge crate washed up on shore. Hauling it above the tide line, you find the crate is entirely intact. You fetch a tool to pry the crate open, and when you look inside, you see . . . a complete Lindal prefabricated wood home. A kit, really; you have to put the home together, but still . . . this sort of bounty is megatreasure from the sea.

One of flotsam's greatest legends belongs to the 1970s, when an Alaskan coastal native did find a crate washed ashore that contained a Lindal prefabricated home and built it for his family. It's a true story, but there's more: Sir Walter Lindal, entrepreneur and owner of Lindal Homes based in Seattle, has reports of several washed-up crates containing his custom-designed prefabricated homes, some fully intact, others in pieces, all from cargo spills. Lindal ships its containers from Port Angeles, Washington, to customers in Alaska and Hawaii. Storms at sea have claimed numerous shipments as the storm-tossed ships lose cargo overboard.

Sir Walter says that more than one lucky beachcomber in Hawaii and Alaska has discovered this high-tech driftwood.

Sir Walter, whose title traces back to preindependence Iceland, was inspired by his ancestors' creative use of flotsam. As a young entrepreneur in the 1930s, he began finding ways to use "waste wood" such as wood flotsam in sophisticated home construction. As his prefabricated-homes business thrived, Sir Walter lived in one of his own designs on a beachfront at Three Tree Point in Washington State. "On several occasions," says Sir Walter, "hunks of our houses washed up on the beach right in front of my home."

In the same decade, a rural commune spontaneously erupted on the Pacific Northwest coast at Shi Shi (pronounced shy-shy) Beach. Back-to-the-landers pitched tents at Shi Shi and lived off the land, illegally hunting berries, mushrooms, and other vegetarian delights found in abundance in nearby lush forests. One couple, we'll call them Dan and Sue, had fled Seattle for the riparian lifestyle at Shi Shi, living in a tent on the beach. As the northbound Alaska current snatches flotsam off the North Pacific Drift, it deposits tons of flotsam on Shi Shi Beach. These currents combine with powerful, often treacherous tidal action off Shi Shi Beach, the result being great heaps of flotsam washing ashore. Since northwestern Washington is logging country, a commercial wood-products mecca with huge cargo ships hauling prime cut logs across the ocean to Japanese builders, much of what washes up at Shi Shi is made of wood. In fact, incoming tides along the northwestern Washington coast are deadly, their waves daily tossing massive logs onto the beaches, scraping them back into the tides and slamming them once again onto the shore. Some of these flying logs that the waves jettison have killed unwary beachcombers. The hippies constructing their squatters' shacks on Shi Shi Beach collected the washed-up logs for building material. (Although surrounded by lush forests, the squatters opposed cutting down living trees.) But hapless Dan and Sue never seemed to catch the big log, and so they vowed to leave Shi Shi before the damp chill of winter set in.

One evening as they sat by their campfire, Dan said to Sue, "This is it, honey. Fall has arrived and we've failed to gather enough wood to build a home. Tomorrow we pull up stakes and head back to Seattle." Sue wasn't ready to give up quite yet. She said, "Let's sleep on it and make the final decision in the morning." Dan reluctantly agreed. Next morning when they stepped out of their tent, they found an entire barge load of precut wood siding washed up at their feet. Dan ran for a hammer and nails.

Dutch Treat

Holland's Terschelling Island is famous for the rich flotsam washing up on its beaches. Like Holland's other barrier islands facing into the North Sea, Terschelling beaches after violent storms receive the detritus of centuries of shipwrecks. In late December 2003, Dutch beachcomber G. Klaase was surveying Terschelling's strandline when he came across a strange piece of flotsam: a wine bottle, some wine inside, shaped like a cannonball with a long neck. Klaase turned to experts who identified it as dating from around 1690. Although it made the news, Dutch flotsamists were only mildly impressed, and no one could convince Klaase to drink the three-centuries-old vino. By a European flotsamist's standards, it wasn't ancient enough to cause a stir.

Flotsam Follies

On the lighter side of organic flotsam, this popular and well-documented tale comes to us from Florence, Oregon. On November 12, 1970, beachcombers discovered the rotting carcass of a gray whale on a beach just south of Florence. Some Firecracker Frank with the Oregon Highway Division, the agency with jurisdiction over coastal beaches, got the bright idea of blowing up the carcass with dynamite. Presumably the next outgoing tide would clear the odoriferous evidence. This made some sense; the agency employed similar tactics to remove boulders that had fallen across highways. Videotaped footage of the actual event is available on the Internet (www.perp.com/whale/index.nc.html) and has forever placed Florence, Oregon, on the oddball's map, for this historic event has developed a global cult following.

Picture a misty day, the tide rolling in. Above the beach, grassy dunes afforded spectator seating for dozens of Florentines who had arrived to witness the dynamiting of the whale. Men scurried to and fro around the forty-five-foot-long, eight-ton carcass. A half-ton of dynamite was placed around the whale, mostly on the leeward side, the theory being that the blast would shoot most of the remains out to sea. In the near distance, gulls could be seen lurking, waiting to score juicy whale entrails. Finally the area around the carcass was cleared. The highway crew beat it to the dunes to join the civilian spectators. Small children trained their eyes on the smelly monstrosity. BOOM!

Smoke rose so thick it obscured the carcass. Then flames shot skyward, and from the dunes came laughing and cheering. Then a woman screamed as boulder-size chunks of whale flesh bombarded the crowd. People ran for their lives. The video-camera's lens got smeared with oily blubber. The lens went dark. A quarter-mile distant, in a parking lot, a chunk of rotting whale meat landed on a car's roof, crushing the passenger side.

The city of Florence and its people will never be the same, for their memories are forever engraved with that one terrifying moment when the sky rained rotting blubber.

On the morning of June 7, 1974, a fisherman off the coast of Trinidad observed a strange object floating on the water near Balandra Bay. The thing, about twelve feet high and fourteen feet in diameter, was terrifyingly similar to images of alien spaceships, those flying saucers from 1950s' black-and-white television dramas. The fisherman contacted police and word soon reached local villagers, many of whom "rushed to the scene while others ran away," according to reporter Phillip Fraser of the *Trinidad Guardian*.

"It was rumored that twenty-four little men—one of them with a radio on his back—had emerged from the strange vehicle and walked inland," the news report said. Brave police confronted the strange floating object and soon identified it as a Brucker Survival Capsule, a sort of pumpkin-shaped survival device made of fiberglass, used in emergencies by the U.S. Coast Guard and many offshore oil operations. Manufactured in LaMesa, California, this capsule contained neither human nor alien bodies, and probably accidentally jettisoned from an oil rig during stormy seas.

Humans have always used oceans as their personal garbage dumps, creating an infinite variety of flotsam and jetsam. The casual strandliner today never knows what might float in on a high tide. Several years ago, Diane Kinman of Bellevue, Washington, was walking along Tillamook Head Beach in Oregon, minding her own business, when she nearly tripped over what at first she thought was driftwood. It had apparently floated in on the last tide. Kinman bent over to inspect the flotsam and came face to leg with a human's prosthesis. Needless to say, the lady was fascinated, and slightly spooked.

The artificial leg included a foot, its toenails painted.

Diane Kinman has entertained herself since then speculating over where the leg came from and what happened to the rest of the person. The leg's skin color was an uncommitted shade of flesh and the toenail polish was true red. Its somewhat battered condition combined with barnacles on the toes suggested it had traveled a considerable distance, perhaps from Alaska on the southbound California current. Or it might have traveled across the Pacific Ocean from Japan, or Polynesia. The artificial leg, wherever it came from, has lately entered the flotsam lexicon: Any piece of beach or tidal flotsam resembling a human body part is today known as a Kinman leg.

Maybe the original Kinman leg was attached to a Captain Hook of the female gender—or persuasion—who was taken prisoner, her artificial leg jettisoned overboard by her captors to thwart an escape. But whoever lost the Kinman leg may still be searching for it, and this truly disturbs Kinman, because she did not pick up the leg and take it home.

Along that sparsely populated, rarely traveled stretch of strandline, the leg most likely got caught up on the next ebb tide and carried back out to sea. Chances are the Kinman leg is still traveling the oceans, enjoying adventures its former owner never dreamed of.

So I said to my psychiatrist, "I think it's time you appreciated the depth of my grieving over the floating stone."

He dangled his Mont Blanc and relaxed his jaw. "You mean," he said, feigning surprise, "you still can't grasp its womb symbolism?"

"I mean, it's time to talk flotsam," I said, irritated. After all, I was paying him.

He surprised me this time, by saying, "What about flotsam?"

Brilliant! For months, I'd been raising the subject of flotsam with virtually everyone I knew, even with strangers. Without fail, the word flotsam seemed to spark a glint, however faint, in a person's eye. Some even waxed enthusiastic, and one actually clapped her hands. "Flotsam and jetsam!" they'd exclaim. "I just love that subject." Yet, as the discussion evolved from Nikes to Legos to yellow rubber ducks, ad nauseam, to the more fascinating and intricate subjects of, say, the six ways a cargo vessel can move simultaneously (heave, yaw, roll, pitch, sway, surge) as it accidentally jettisons flotsam-to-be, or the fetch of a wave and the reach of its fingers, an ocular veil would glaze the person's eyes, followed by a yawn, and then, inevitably, a change of subject. The only folks who stayed alert were those with their own flotsam stories, and some of these were doozies.

Kat Silverman, for example, a Britisher who lived for many years along Cornwall's seacoast, immediately lit up when I mentioned flotsam. Kat has flotsam memories not only of the many shipwrecks along the Cornish coast; she also recalls a particular incident in which some local men discovered a large vat of scotch washed up on a beach not far from Sharpnose Point. Because the vat was too heavy to move off the beach, the men organized their drinking fest right there. Kat could not recall which single-malt distillery had lost its vat; nonetheless, I've lately considered relocating to Cornwall.

A stranger from Perth, Australia, responding to my mention of flotsam, recounted his discovery of a human hand washed up very near his beach cottage. He fetched the nasty flotsam off the strand, marched straight to the local police department and, well, handed it over to a duty officer. Upon investigation it was averred that the macabre flotsam had belonged to a sailor aboard a pleasure craft who had apparently been tipsy when he tried slicing himself a wedge of cheese or sausage or some such snack at the exact moment his sail boat heaved, yawed, rolled, pitched, swayed, and/or, surged.

Almost everyone has a flotsam tale, some more fantastic than others. And while the amateur beachcomber possesses only one or two really good flotsam tales, professional flotsamists can carry on for hours about their intertidal discoveries. In my quest to delineate between a beachcomber and a true flotsamist, I found that the vast majority of truly fascinating stories come from casual beachcombers, folks who don't go out looking for flotsam but who stumble upon such wonderful treasures as a vat of single-malt. Flotsamists have many tales, but on average, they are somewhat less enchanting. It's the accumulative quality of their pastime that makes a great story.

And the flotsamist possesses special personality traits. This was the subject I wished to discuss with my psychiatrist.

"These folks are nuts," I started.

The doctor flashed a cautioning hand. "Pot calling a kettle black," he said.

"They are obsessed with flotsam," I said, avoiding his pitiful attempt to divert the subject. "I heard about this so-called beachcomber in Orange Beach, Alabama. You know, the place at the center of that hurricane."

"Which one?"

"Whichever hit Orange Beach. So this beachcomber guy, I'm told, is collecting disaster souvenirs. So I track him down, and sure enough, the guy's a hardcore flotsamist. Has been for something like forty years. But, guess what? Are you listening?"

He nods, half-comatose.

"The guy only collects evidence that supports his theory of alien landings in the Gulf of Mexico."

"You mean boat people?"

"I mean 'alien,' as from Mars."

The shrink straightened up in his wing chair, coughed lightly, jerked his chin out the way men do when their ties are choking. He swung the Mont Blanc like a metronome and said, "Time's up."

"I'm talking obsession here," I push. "Flotsamists are absolutely fixated on collecting stuff. Alien spaceship parts, whatever. They're totally obsessive-compulsive. Now, I know what you're thinking. You're thinking about my floating stone. But that's not obsession. I am merely . . ."

"I'll see you on Thursday," he said. "Meanwhile, I suggest you stop fixating upon that little floating stone and invest the attention in a new pastime."

But it was no use; flotsam had become my life, fitting me like a second skin. I spent more and more time in the water, chasing flotsam, diving for lagan. When I wasn't in the Sound, I was standing on my deck with binoculars, scanning the beach for freshly washed-up objects, a habit that preceded daily beachcombing forays. As the kitchen counters and library shelves filled up with flotsam discoveries, the house gradually acquired the pungent scent of fresh wrack. The deck and living room were merely an extension of the beach, and seabirds thought nothing of crossing the threshold. I passed a summer battling sand crabs and tripping over rusted ship parts I'd salvaged from the sea floor. Alas, like a tunnel-eyed bottom scavenger, I lacked a broad perspective.

Unlike their primitive ancestors, twenty-first-century beachcombers don't scour beaches for food staples, goiter cures, and construction materials. Today's beachcomber is more often a strandliner of leisure, clad in chic beachwear—or else American—searching for the cheap thrill of it, finding something for nothing, or an individual who believes Neptune has lobbed a gift especially to him or her, material or spiritual, viz., the vision-questing lady, the more remarkable the better. Alas, today's flotsam, jetsam, and lagan is 99.44/100 percent debris, and the .56 percent worth a eureka is, ironically, usually plucked off the waves by novice beachcombers or the casual strandliner plaguing the beach purely for its palliative effects. These juttering interlopers can't appreciate what they've stumbled upon, and usually after mulling it over, lob the treasure back into the sea from whence it came. This sort of beach bum is the flotsamist's cruelest thorn-in-the-side nemesis, a creature simultaneously to abjure and to stalk—for his ignorance is the flotsamist's reward.

The flotsamist finds intrinsic value, or scientific significance, obvious or potential beauty, and sometimes even aesthetic properties in objects ordinary beachcombers find ugly, uninteresting, or even blots on the environment. The beachcomber stumbles upon a barnacle-coated, rusty, unidentifiable length of steel and curses it, or just ignores it. The flotsamist gathers up the object, like so much golden fleece, takes it home to ponder over, to find a practical use for, or just to have around to look at. With Sherlockian devotion, the flotsamist will spend weeks, months, even years researching an object's point of origin, tracing its voyage across the oceans.

The world is awash with these oddballs; and flotsam eccentrics ply the tide lines of every country embracing coastal waters. This universal flotsam fetish is nowhere more pronounced than in the northernmost islands of the Netherlands. Take, for example the strandliners of Texel and their flotsam museum.

The price of admission is worth every kroner. Go Dutch or pay for your date, either way, a visit to the Jutters (Beachcombers) Museum in Texel, the Netherlands, will send thrills up your favorite flotsamist's spine. The museum explains: "Beachcombing is second nature to the people of

Texel. For centuries, it was a necessity. Wood was always needed to stoke the stove and build with. The cargo vessels stranded on the coast of Texel were often a supplement to the local people's frugal existence."

The *strandovers* of Texel, like beachcombers everywhere, love nothing more than a great flotsam discovery, except, perhaps, telling the flotsam tale. In Texel's Jutters Museum, flotsam collected off area beaches is displayed—where else?—in the Jutterii (strandover's museum), where every day from 11:30 a.m. to 1:30 p.m., Texel's strandovers and shipwreckers and their fans gather there to swap flotsam tales, each focused on a particular item of flotsam found on Dutch beaches.

In 1931 the Netherlands passed the "Jetsam law" prohibiting beachcombing on its beaches. Whatever washes up in Holland belongs to the state unless its owner can prove right of possession. Each case is settled in the locality where the flotsam is found. The local mayor is also chief wreckmaster of beaches within his or her jurisdiction. Texel's mayor has six deputy wreckmasters who beachcomb for flotsam along the coastlines of the North and Wadden seas. Still, illicit beachcombing is commonly practiced, particularly for driftwood to use in construction.

Texel is one of the Wadden Islands; it's the northernmost island facing into the North Sea, with the Wadden Sea on its east-southeast shores. Thousands of ships have wrecked off the storm-wracked North Sea and Wadden coasts. On a single night, Christmas Eve 1593, nearly two hundred ships, unfortunately mostly lashed together, went down during a violent storm. Today, ships commonly strand or go down along Texel's coastline. The Jutters Museum is home to one of the world's finest messages-in-bottles collections: Cor Ellen's bottled mail collection, circa 1950–2003.

Seashell Desecration

Today's world of crafts is rife with cheesy seashell-encrusted offerings—seashell-encrusted lamp shades, cigarette lighters, drink coasters, placemats, picture frames, jewelry boxes, bathroom tissue holders, even computer monitor frames, for criminy sake—all of which should be bought for the single purpose of dismantling them before someone gets the eureka to place one inside a time capsule. One craft teacher instructs her students to boil seashells they find on beaches, to ensure the animal inside won't stink up the *tchotchke*. What glue are these crafters sniffing?

Flotsam art is an extremely complex genre, not suitable to the crafter. In fact, North America's Van Gogh of flotsam art, Jay Critchley of Provincetown, Massachusetts, never works in shells; Critchley, a true artist, sometimes works in beach whistles, as will be described eventually. As for flotsam crafters, none are represented in the American Museum of Folk Art, and if I were a seashell, my worst nightmare would be spending my legacy Krazy-glued to a black velveteen tissue-paper box. On the other hand, I own a Victorian snuffbox fashioned from two limpet shells edged in gold banding with a handmade gold clasp and hinge, and I defy anyone to sniff at its extraordinary beauty.

Seashells, in their original unglued, pre-tchotchke state, can be worth thousands of dollars. Indeed, many Pacific Islanders, Filipinos, Japanese, and coastal Africans work fulltime as beachcombers or undersea divers, seeking the shell equivalent of the Golden Fleece. Since the early eighteenth century, Bohol, on Panglao Island in the Philippines, has been a center of the profitable seashell trade. Local markets hawk to private collectors and seashell businesses, many of whom today resell them on eBay. Buyers from all over the world visit the Bohol markets, and competition for rare specimens is fierce—even tinged with foreign intrigue, in some cases, the stuff of crime thrillers.

Bohol isn't alone in the shell game. Throughout the world, divers illegally harvest seashells to supply a thriving international black market. It's possible to pay upwards of $50,000 U.S. for a particularly rare and coveted shell in excellent condition. Fishermen often illegally sell shells caught up in their nets to wholesalers. Harvesting living animals has caused some species to vanish, the price a species pays for its exotic beauty. At least these collectors aren't gluing them to clocks and cookie jars. At least, I don't think they are, but you can't predict what devil-may-care trends are infecting the nouveau riche.

Beach treasures that crafters covet include the sand dollar and the starfish. These hapless creatures washed up onto a beach—some still alive—form the nexus for some of beachcombing's rowdiest fisticuffs. Serious snatching competitions break out, epithets are muttered or even shouted; I once overheard a California beachcomber say to another, "I don't care if you got here first, I saw that sand dollar through my binoculars from fifty yards down the beach. It's mine by first sight."

First-sight beachcombers never attain the rank of flotsamist. A flotsamist understands possession is nine-tenths of the law; touch it and it's yours. Only for heaven's sake, leave the live sand dollars and starfish alone. They'd rather die from exposure to sun than to a crafter's glue gun.

The Amazing Captain Crowe

Because my heritage traces back through five English sea captains, and because I grew up fending off boyfriends' advances beneath the misty tracking eyes of my great-great-great-great-great-great-grandfather Peter's sea captain portrait, I think I know what a man of the sea is supposed to look like. Adding to my credentials, I've met a lot of ship's captains around the world, some, like their vessels, more seaworthy than others. The Spanish captain of a Morocco-bound steamer who let me take the ship's wheel as he delicately balanced his hand on my waist later delivered us safely from the mawing jaws of a freak Mediterranean storm. He ranks among my favorite sea captains, not only for his graceful hand and maritime expertise, but also because Captain A-hem was an expert on Mediterranean flotsam. In fact, I learned from Captain A-hem that a popular jetsam flung overboard off the French Riviera is clothing, most commonly ladies' bikini tops, and usually haute couture. The captain recalled for me a particular wire-cupped size 110D (U.S. size 42D) Aubade *soutien-gorge* he'd personally retrieved off an incoming Riviera tide. He even offered to let me try it on, and when I quipped I'd only need half of it, he lost interest, and I never got to view his collection. Some flotsam is too hot to handle.

Off the Latvian coast, during the Soviet occupation of the Baltics, I hung out with Captain Janus Vitols on his oceangoing fishing boat. Our common language was Spanish, convenient since Natasha the Russian journalist dogging my heels didn't understand it. Captain Vitols has plied the world's oceans and braved major storms—the sea is chiseled into his jaw, and though he has frequently fished off the Chilean coast, he's never encountered the Chilean Blob. I wondered about his experiences with *pecios a la mer*. The chiseled jaw tightened. Captain V. said, "*Yo me busca bastante tipo raro a inundar la calles de Moscu*," or, loosely translated, "I've caught enough queer fish to flood the streets of Moscow," a cynical political jibe he knows the loyal Soviet, Natasha, cannot translate.

So when I say that I know what a ship's captain should look like, I think folks should sit up and take note: Captain John Crowe of Newport Beach, Oregon, is Neptune personified. Not only is Captain John one of the world's most accomplished deep-sea divers, he also possesses a heck of a flotsam and lagan collection, all of it hauled up from the sea floor or plucked off the ocean waves by Captain John himself.

Professional divers like Captain John work jobs that take them deep beneath the ocean's surface, where they may have to perform intricate repairs on ship's propellers, or check a hull for leaks, or rescue a person or a valuable item that accidentally plopped into the drink. This true man of the sea has seen more flotsam, jetsam, and lagan than all of the Baltic populations combined.

Captain John Crowe was born in a Southern California beach town and spent his childhood swimming and snorkeling off Laguna Beach. As a young boy, he was one of Catalina Island's "little imps," as they were known, the boys who met the tourist boats sailing into Catalina's Avalon Harbor. Tourists would toss coins into the sea and the boys would dive off the rocks, recover the coins underwater, and resurface, holding the coins aloft to the amazement and applause of the tourists. As a schoolboy, John Crowe regularly skipped school with his beach buddies, swimming and snorkeling off Laguna Beach at a spot called the Thousand Stairs. The stairs ran down a steep vertical cliff to the beach, "not really

Captian John Crowe at the flotsam portal to his home.

a thousand of them, but enough to keep the truant officers from coming after us." At fifteen, Captain John took up diving, and he hasn't stopped in more than fifty years.

Over a lifetime career, Captain John has dived in the world's oceans for NOAA on research vessels and for other government agencies and private contractors. He possesses finely honed underwater instincts he's had to depend on long before the advent of technologically sophisticated diving equipment. This keeps him in demand; when a really difficult dive is planned, it's John Crowe the experts call in to do the underwater work. When John needs assistants, he looks for surfers. "They make good divers," he says.

Today, Captain John lives with his wife, Patty, in Newport, Oregon, where he bases his diving company. The Crowes' front drive, its amassed flotsam collection forming a wondrous barricade against landlubbers, informs even the casual passerby that true mariners live here. The front gate is a ship's interior door complete with porthole, and on either side of the door, lush gardens are landscaped with giant fish-net floats that dwarf Pilates exercise balls, a propeller that could have driven the Titanic, a massive rusted anchor, ships' lanterns, bells, giant buoys—name your flotsam, it's in Captain John's front yard, and every object was fetched from the sea during his underwater exploits.

Diving extracts its toll on even the strongest human, and so not long ago, Captain John was hospitalized for heart bypass surgery. As is usual in bypass surgery, surgeons remove sections of arterial vein from the patient's leg to replace the heart's damaged arterial veins. Following the successful surgery, Captain John's leg became horribly infected.

"They took a section of my leg the size of a two-by-four," says Captain John. "From up near the groin all the way down my leg; they cut it all out." But the infection blossomed and physicians began talking about amputating his leg. Captain John made an appointment with his surgeon in Corvallis. The day before the amputation surgery, the surgeon died, leaving Captain John in a hospital with the leg infection threatening to kill him.

"I wouldn't even look at it," he told me. "My wife did, but I never looked at it. They said it looked horrible, and I could smell the infection. It smelled awful."

A physician stepped in for the deceased surgeon and after trying everything he could think of, he prepared to amputate the leg. Then a nurse took the doctor aside and suggested he try seaweed. Most American surgeons balk at any suggestion from a nurse, let alone what sounds a lot like a folk cure. But this Mexican physician went with the idea, ordered some sterilized seaweed, and packed Captain John's leg with seaweed (Note: Do not try this at home). And waited. The cure didn't take long, and today Captain John walks gratefully on his own two feet, still operates his diving business, and continues retrieving some of the world's biggest flotsam from the ocean waves.

Off the Wrack: Seaweed Flotsam Revisited

Flotsamists love nothing more than a good wrack. It's that green-to-brown seaweed and kelp line that forms on most saltwater beaches at high tide. The thicker and fresher a wrack, the more its attraction to a flotsamist, for within that twisted, braided, mashed, and crumpled mass are found the jewels of flotsaming. My little pink plastic propeller was rescued from wrack, as were several excellent plastic fishing bobs and perhaps a third of my porcelain shard collection. You never know what a good wrack will contain unless you're willing to crouch down and get personal with it. The serious flotsamist pays no heed to smirking beachcombers who pass by in search of, you guessed it, the Golden Fleece, for within the wrack lie uncountable exotic treasures . . . but wait: The wrack itself floated in on the tide, and is flotsam of the first order.

Today, seaweed is commercially marketed as an antioxidant, which may prevent damage to the human body caused by cancer and by environmental toxins such as heavy metals and radiation by-products. Scientists at McGill University in Montreal have shown that a product derived from the algae wakame binds to radioactive strontium 90 in the human body, allowing the toxin to be excreted with the binding material. Some physicians recommend ingesting a wakame-based product before having X-rays, in the belief that any leaking radioactivity from the X-ray will be eliminated from the body along with the wakame.

Our distant ancestors' cure-all has lately enjoyed popularity both for its cosmetic benefits and healing properties, John Crowe being a fine example. Taking a cue from Eastern medicine, surgeons in the West have lately been applying sterilized seaweed as a topical treatment to reduce swelling at incisions and for improving skin texture. Meanwhile, in the era of über-haute cuisine, chefs around the world prepare entire meals using only sea-derived foods. Arame, harvested on the beaches of the Ise Peninsula on Japan's east coast, has for more than a thousand years been used as a sacred offering.

This organic flotsam is harvested in late summer at low tide, when the plants are still young and tender.

Ever looked inside kelp? First thing you notice is its thickness. Kelp is another organic wonder that washes up on saltwater beaches. Known for its high nutritive value and mineral content, kelp is sold as vitamin supplements, tea, and compresses for arthritic joints. As a food, kelp is served raw or cooked, and my one childhood kelp-as-food experience reminded me of chewing on salty rubber bands. The sea plant is even hawked as a cure for cellulite and for weight loss, but don't waste your money. Finally, kelp is considered an effective diuretic. Next thing we'll hear about is kelp catheters.

Today's consumer may find marine-based cosmeceuticals and body therapies at their local drugstore or order them from the squealing Shopping Channel ladies whose inflated testimonials cause grown men and women to weep and spend.

I've had a seaweed body wrap. To be perfectly honest, I felt like a sushi roll. The afterglow, though, started me wondering: What if I went down to the beach in front of my house, gathered seaweed from the wrack, and made my own seaweed wrap? I could save $250, not counting tip.

Several days after the alginic eureka, a particularly rich wrack formed on the beach. I slipped into my skimpiest two-piece bikini and aqua socks and tripped down to the wrack, bent over, and lifted a slimy green ribbon of seaweed. I felt the supple collagenic nature of the plant, which causes skin to plump out when it's applied. I felt the zest, a pleasant sting that awakens the flesh on contact. I thought how much seaweed was out there, hundreds of varieties, how the mounds of seaweed continually pile up along the tide line, at the very least, a lifetime supply of body wraps, facials, body creams. Nature is generous, I mused, staring into the wrack.

It was early morning. The summer sun had not yet risen high enough to heat the wrack and dry it up. I looked around. Two beach visitors clutching Starbucks cups like lifelines sat chatting on a nearby log; otherwise, the beach was deserted. I stared down the Starbucks couple until they got annoyed enough or scared enough to leave the beach. Alone now, I lay down, positioned myself on the damp seaweed, and gripping one end of the wrack to my waist, rolled over until my body, shoulders to knees, was encased like a rice ball in heaven.

Almost instantly my body enjoyed a stinging sensation. Ah, the stuff's working, I thought. The thrill of proving my wrack-wrap theory drove a chill along my spine. I could launch a seaweed flotsam revolution; everyone would learn to appreciate a good wrack. Soon the world's beaches would be cheek to cheek wrack wrappers. But wait a sec. I'm a loner, don't like sharing the beach with anyone. Better keep this secret, I told myself, and, turning over, I noticed a Camel cigarette pack tangled in my décolletage, then the oily sheen of some engine fuel—or was it tar?—clinging to my arm. Barnacles, little sticks of splintered wood, sand, and, yes, it was tar. I sat up, and that's when I saw the divers who had surfaced off the tide line—the local diving school trains in front of my house. One of them had spotted me and gestured to the others. Now they waved. I pretended not to see them as I peeled off my seaweed overcoat. What for criminy sake was I thinking? Puget Sound is seething with pollutants, not to mention diving students.

Swimming Pigs Meet the Big Wooden Phallus in Emperor Ishii's Office

Japanese are known—even revered—for their culture's perceived eccentricities. Actually, I believe Japanese are simply better at executing eccentric affect than other folks. Japan's anime far outstrips any similar artistic effort tried in the West. Tokyo, so technologically advanced, is so creatively electrified that no nuance escapes its luminous fingers or the artist's lightning-bolt interpretation. Tokyo is anime: outsized, stylized, colorful, and futuristic, edged in sex and violence, pulsing, pounding, driven.

The world of anime seems far-flung from the world of Flotsam Emperor Tadashi Ishii and his passion for beachcombing. But that's only if you haven't ever been immersed in Ishii lore. Ishii's passion for his subject is outsized and colorful. What he discovers along the beaches of his country often are remnants of violence or strife, occasionally nuanced with a sexual significance the polite man chastely refers to as Yin and Yang. On Japan's beaches, Ishii inevitably discovers fascinating trinkets, many speaking of humankind's foibles and nature's freaks. And Ishii navigates the world of flotsam as a loner, appearing at an instant on beaches like flashes of magic light dancing off the glistening tide. If Ishii missed it, it isn't there. If you blink, you'll miss Ishii.

Ishii has collected hundreds of wooden objects off Asian beaches—among the more fascinating, three carved phalluses. One phallic reproduction is small and thick. The second, measuring a healthy fifteen centimeters, is made of lauan wood and washed up on a Fukuoka beach in southwest Kyushu. The largest phallus Ishii has so far found measures twenty-one centimeters (a bit over eight inches) and is made of Japanese cedar. This big guy Ishii plucked off the tide at Eguchi Beach.

Usually made of stone, phallic fetishes date back as far as ancient Greece. But carved wood phalluses have for centuries been attributed to Tantric and Buddhist sects, including worshipers of the Thai Goddess Tuptim, whose shrine in Bangkok is unique for its erotically charged Tantric-style influence. Chao Mae Tuptim is a fertility goddess, and her shrine is visited by women wishing to bear children. A sign at the shrine explains that Tuptim has received thousands of phallus-shaped gifts "both small and large, stylized and highly realistic. Over the years they have been brought by the thousands, and today, fill the area around the shrine."

Virtually nonexistent as a fetish symbol in other East Asian Buddhist sects, the wood phalluses may likely have traveled ocean currents via the Gulf of Thailand, northeast into the South China Sea, northward into the East China Sea, and branched off into the Tsushima Current to wash up at Eguchi and Fukuoka. Ishii doesn't pretend to know the stories of the three washed-up phalluses, and neither do I, but oh, what fun to speculate.

In December 1966, Ishii found a root carving of a sea lion washed up onto a Fukuoka beach. He has also found dolls, many made of wood. Ishii explains that there are two kinds of doll flotsam, toy dolls and religious dolls. On February 11, 1981, Ishii was walking on the beach at Shiraishi when he found a doll's body with religious clothing wrapped around it and Buddhist scriptures on the body. A few minutes later, he found another, then another, for a total of ten.

Ishii said he found the experience scary, but that's not the scariest that things got for the Emperor of Flotsam.

In April 1996, Ishii found a severed wooden head washed up at a beach in Fukuoka, an especially scary find, because a few days earlier, a beautician had been murdered and the killer had scattered her body parts around the prefecture. The severed head that rolled ashore gave Ishii the willies, but there's more: In June 1997, in an incident that sent shivers around the world, a Japanese junior high school student methodically cut off the head of a young boy, then placed the head on a wall near the school. That same day, at a beach not far from the murder scene, Ishii found yet another wooden head washed up. Ishii wanted to photograph the head for his flotsam cataloging, but his family, fear some indefinable yet palpable repercussion, wouldn't permit it.

Ishii says he doesn't much enjoy finding flotsam heads, or dolls. But Buddhas are something else, and he's found more than his share of the Enlightened One upended and meditative at water's edge. On December 7, 1979, at Katsuuma, Ishii found a carved wood Buddha. From its condition, Ishii concluded the old statue had traveled numerous times around the North Pacific Ocean, riding surface currents that bleed into each other. Ishii compares the Buddha to the wandering monk who constantly travels, begging his rice and sake. The old wood Buddha appeared to have traveled the ocean for decades, if not hundreds of years, before floating ashore at Katsuura. Ishii says old Buddha statues aren't usually thrown away, but are returned to their temple or shrine, or—and this may explain this old fellow—they are brought to the seashore, placed in a small boat, and set adrift.

Today the old Sailor Buddha sits in Ishii's office so that whenever Ishii feels a need to petition, Buddha's handy.

Among his other wood findings, Ishii once found a box that had washed up on the beach in northern Nigata Prefecture. The box was lettered in Russian and had presumably traveled across the Sea of Okhotsk from Russian Siberia to Japan. The lettering identified it as a "bomb box." No bomb inside. Because so much Russian flotsam arrives on Japan's northern beaches, Ishii believes the post–Soviet Russian population chucks all the old propaganda into the sea.

Another among Ishii's favorite finds is a toy horse made of *yoki-zukuri*, or particle wood. The horse washed up at Kitahara seashore. Where it came from Ishii couldn't say for sure, but one thing was certain: The sea gods had sent the Emperor of Flotsam a horse—an auspicious sign.

Ishii also has collected a multitude of plastic toys. Among his most interesting are the reproductions of anime heroes, usually made in Taiwan, and a collection of military-themed toys from North Korea, including soldiers with parachutes, tanks, and guns. In 1996, Ishii found several plastic piggy banks washed up on a Taiwan beach. Blue, yellow, or pink, the piggies ranged in size from nine to thirty centimeters. Ishii isn't the only one finding pigs on the beach: In September 2002, author and flotsamist Ed Perry found a yellow piggy bank washed up on a Florida beach. Another piggy bank, this one green, was found by Florida beachcombers. Then there's the cigar-smoking piggy with the bowler hat plucked off Florida's Sebastian Inlet by Robert Nordstrom. Based on a manufacturer's code on the pork belly, Nordstrom's five-inch-long smoking-swine savings bank may have been manufactured in the Dominican Republic by the now defunct cigar maker Sabrosito.

Pursuing the smoking theme, let's consider the 100 Yen lighter, a brand of cigarette lighter that's like a skinny Bic with ribs, usually clear green or blue plastic and about eleven and a half centimeters long (just over four and a half inches). During my early China days, at any given time, I owned half a dozen 100 Yen lighters, because back then I smoked. I coveted a red 100 Yen; they're scarce. In fact it may have been a red 100 Yen I lobbed overboard on a trip down the Yangtze shortly before I kicked the habit during a hellish two-week withdrawal in Hong Kong.

The 100 Yen is ubiquitous; it seems to clone itself in China's streets and alleys, and even on the beach. I have found numerous 100 Yen lighters of the common variety washed up on Hong Kong's back harbor on the China Sea, and on the beach at Macau, where even lungfish choke.

Ishii has a huge collection of 100 Yen lighters. In his flotsam encyclopedia, he explains there are two sizes of 100 Yen lighters: the ones that are eleven and a half centimeters long, and a less common model that's just over eight centimeters, presumably for toddlers' hands. When searching an agricultural drainpipe at Yazikimachi, Ishii found sixty of the lighters. Five were from Korea; twenty-five were green, eight were blue, and—wait a minute—none were Chinese.

While flotsaming at the Shiraishi seashore in November 1990, Ishii found a plastic *doruharubana* washed in from Korea. Ordinarily carved from stone, doruharubana are statues placed in villages to protect residents from disease. The nine-centimeter pink plastic statue caused such a sensation in Japan that a television crew covered the event. Then, in 1994, Ishii found another plastic doruharubana, this one black, washed up at Japan's Katsuura seashore.

Ishii, the scientist, meticulously maintains records of his finds, including date, time, and location, the information critical to its value as evidence of ocean current shifts and weather patterns. Among his more spectacular finds is a commemorative bowl made of silver and engraved in gold, which Ishii found washed up on a beach on Fukuoka's east side. The bowl, dated 1923-1924, was a presentation piece celebrating the warship *Izumo*'s first circumnavigation of the globe. The *Izumo* was built by British shipbuilding company Armstrong in 1897 and launched in 1899. What I want to know is why anyone would jettison such a keepsake into the sea. My romantic impulse suggests it must have gone down with a ship.

Dragon Lagan

Tadashi Ishii is perhaps proudest of his collection of more than four hundred ceramic, porcelain, and celadon bowls. He's found many ancient rice bowls and vegetable bowls of varying sizes and decoration, some from the Edo or Tokugawa period (1600–1868) in Japan. Besides these incredible treasures, Ishii has collected numerous *nikkeisui* bottles—decorative vases from the Meiji period (1868–1912). Nikkeisui, extremely rare flotsam finds even on Japan beaches, are made of handblown glass, blown into three or four connected global tiers with simple long beaks, and may have been used in ikebana. The rarest nikkeisui Ishii has found is a six-tiered vase.

Having invented pottery making around 4000 BC, the Chinese also hold the record for the most pottery flotsam and jetsam. When in the fifteenth century AD, Spanish galleons began hauling China's pottery urns, dishes, decorative pieces, and rice wine jugs around the oceans, the seas swallowed up thousands of tons of pottery that went down with wrecked ships or were jettisoned in storms. Packed in Spanish Manila trade ships bound for New Spain (Mexico), some intact urns and millions of pottery shards began beaching along North and South America's coastlines. Some urns were etched with Chinese characters. Meantime, massive jugs of Spanish wine went down with ships in hurricanes off both American coasts.

A few lucky American beachcombers have recently stumbled across great pottery urns that rolled ashore between 2001 and 2005 along the U.S. West Coast. Some of these have been exhibited at beachcombers fairs at Long Beach and Grayland in Washington State, including the mysterious unglazed stoneware urn found by Steve Sypher two miles north of "beachcomber's heaven," Tokeland, Washington. When he found it, the urn wore a light barnacle coating, an indication it had not traveled for a long time in the ocean, and yet it is probably the most primitive urn of the eighty or so urns so far found along the West Coast.

The oblong, slightly buoy-shaped urn stands nearly two feet high and is just over sixteen inches in diameter at its broadest section, very narrow at the bottom, its opening apparently plugged with clay. The light brown urn is etched with primitive diagonal lines across most of its girth. Its narrow bottom is etched with circular bands. It's empty, although

Sypher says it rattles as if a small stone or pebble has been caught inside. When I saw this object at Grayland's forty-first annual driftwood show, where it claimed first place, I was told Sypher had found the urn only on the previous day, May 14, 2003. Since then, several more urns have been reported washed up along the Washington coast, although none as ancient looking as this.

Some flotsamists single-mindedly collect only ceramic originating in Asia, shards of broken objects found along tide lines. These are particularly abundant on Pacific Rim coastlines. A pottery shard can be dated by the formula of clay, porcelain, or celadon used to make it, or by the decorative images on it. While Asian beaches receive the vast majority of pottery flotsam, the U.S. West Coast and Alaska also attract these tiny shards of history, each holding a memory of its creator and a question about how it came to be flotsam. But a pottery flotsamist quickly learns that most of what washes up in exquisite little shards is either recently manufactured or the work of clever antiquarian pottery forgers. China daily pours out millions of clay, porcelain, and celadon reproductions, which are pawned off to tourists who are under the impression they are acquiring genuine antiques. No one fakes antiques better than the Chinese.

In 2001 a Chinese diving team of twelve underwater archaeologists visited the remains of a thousand-year-old shipwreck. Dubbed, "Shipwreck No. 1 in the South China Sea," the news media are fond of calling it "China's *Titanic*." The team salvaging the site works hard to keep the shipwreck out of the news. That's because Shipwreck No. 1, a hundred-foot-long oceangoing wooden vessel, contains as many as sixty thousand relics from the Song Dynasty (960–1279). No doubt during the thousand or so years the wreck has lain on the sea floor, some of the cargo broke loose and went adrift in the currents to wash up on beaches around the region, if not around the world. Already the salvagers have recovered more than ten thousand items, including priceless ceramics and porcelain bowls. Among the ceramics salvaged were a large number fashioned in distinctively Arabian styles. Based on this evidence, and on the history of merchant vessels traveling from China throughout the world, the team thinks the ship was en route to the Middle East when it went down, likely in a typhoon.

More important for flotsamists, the South China Sea is believed to hold the remains of more than two thousand shipwrecks. Officials at the government's South China Sea Research Center based in Hakou, Hainan Province, worry that the amazing treasures lying at the bottom of the sea risk "being damaged or endangered by mushrooming illegal salvages and ensuing dealings in the international art markets," says the center's director, Wu Sicun.

He Shuzhong, a cultural heritage government official, worries because "the world's treasure hunters have turned their attention to ancient sunken vessels in the past decade. They can often make greater profits with fewer risks out of the commercial salvages than out of tomb raiding." He refers to a massive salvaging of shipwrecks in the South China Sea by one British salvager, who racked up $20 million at a Christie's auction in Amsterdam selling the ceramics and gold from a Chinese ship that sank in 1752. The auction was in 1986. A year later, China began protecting its ancient lagan. One official remarked, "In art markets of major Chinese cities we can often see antiques with shells clinging to their surface. They have been taken from shipwrecks."

Not necessarily so. As the informed flotsamist knows, a clever antique dealer can make anything look old, including flotsam; gluing seashells to a fraudulent flotsam or lagan find is a piece of cake. Just ask the shell seekers who have wrestled with barnacle glue.

Flotsamists will be interested to learn that Shipwreck No. 1 was discovered just a hundred feet offshore at a river mouth, where presumably the placement and angle of the wreck combined with the force of the river's flow helped keep the boat from being buried too deeply in sand. However, if a flotsamist plans to violate international law, or perhaps take advantage of its gray areas on salvaging undersea treasure, he or she would best refrain from imitating the bands of Chinese fishermen looking to get rich: They bomb the shipwreck sites with explosives, retrieving whatever's left intact.

Pumpkins, Sausages, and Rolling Pins

The world's serious flotsamists have their special beaches, and these are based on the ocean currents, tidal action, weather, and sundry personal tastes. Lieutenant Colonel A. H. Richardson, U.S. Marine Corps, travels once or twice a year from his Maryland home to Japan to beachcomb for glass fish-net floats. Richardson's house is a colorful jungle of these jewel-like bubbles, all of which he harvested, and in the blink of an eye he can tell a genuine glass float from a fake. He may be one of the West's experts on Japan's beach flotsam, and like other devoted flotsamists, he has his favorite beaches. But flotsam doesn't select its beaches the way people do.

The posh beach communities of Laguna Beach or Martha's Vineyard may regard their strands exempt from receiving a headless corpse or a couple hundred Nikes—all left-footed—or a half ton of putrid, decaying *Velella velella*, but in all its organic and inorganic wonder, flotsam goes where it goes. Winds and currents deliver flotsam to its destiny. The surprise is, as Richardson says, just another reward of the hunt.

Other collectors of Asian glass floats have established places in flotsam history. Alan Rammer, marine education specialist for the Washington Department of Fish and Wildlife, is a glass-float specialist who often lectures on the subject. Rammer, assisting his late mentor, the flotsam author Amos Wood, were first among members of the world flotsam community to develop a systematized protocol for classifying glass floats. Their work inspired today's most renowned glass-ball collector and specialist, the cigar-chomping, eagle-eyed Walt Pich, whose book *Glass Ball* is a kind of bible for the subgenre of Asian glass-float collecting.

Pich, who also wrote *The Beachcomber's Guide to the Pacific Northwest*, has been collecting glass floats ever since he saw his first one roll up on a California beach. The pastime turned into a serious hobby, and as Pich amassed thousands of glass floats of every description, he began traveling to Japan to research the history of the Japanese glass fishing float. This research forms the basis for the book *Glass Ball*, although Pich hastens to credit his predecessors, notably Amos Wood, and says his own classification system only builds on Wood's work. But Pich, too, has contributed enormously to the science of glass-float identification, and flotsam professionals depend on Pich's guide to classify their finds.

Pich's book describes the difference between a doughnut-mold jumbo glass float and a dot-mold jumbo. "The doughnut mold is often found with an appealing blue cast," he explains. "It may be the oldest of the jumbo rolling pins." Pich describes the various characteristics of floats that lucky beachcombers find on the tide line, including size, color, swirls, manufacturing techniques, flaws, etc. Through use of photos, diagrams, and Japanese characters, he deciphers a float's identity and gauges its value on the market—a market that is huge and profitable.

The "pumpkin float," a relatively new bit of flotsam, is named, like all glass floats, for its shape and may be one of the rarer glass floats found washed up on the U.S. Pacific Coast. Pich devotes an entire chapter to it, including the intriguing mystery surrounding the pumpkin's origin. The sausage float is another specimen altogether. Shaped like sausages, they are old and Japanese, Pich says, and are interchangeable with spherical floats.

At a 2005 beachcombers fair in Ocean Shores, Washington, I was talking with Pich when a man approached Pich's booth. "Did you get it?" Pich asked him. The man shook his head. "Naw," he said. "Somebody told the guy that his float was worth probably ten thousand, and when the guy heard it, he clutched onto his float and ran out the door."

Fakes abound. I bought one myself from a booth at the fair. I had admired the unusual blue cast of the float's glass, similar to the color of a Bombay Sapphire gin bottle. The booth operator explained that this was a rare Japanese float that he'd found washed up on the Northern California coast. Rare cost me ten dollars. Pich later verified the float was a phony repro, which I had suspected based on the price, but I'd bought it anyway because it reminded me of martinis. Had I read Pich's book before making the purchase, though, I could have enjoyed the satisfaction of calling the merchant on his dupe.

Master of the floating glass ball, author Walt Pich, with one of his larger specimens.

Norwegians are credited with inventing the glass fishing float sometime before 1840. Credit is sometimes given to Christopher Faye of Bergen, who in 1865 won a gold medal for his invention: blown-glass balls, five or six inches in diameter, for use as fish-net floats. These first creations were small egg-shaped affairs tied to a single fishing line and hook. Glass floats were buoyant and didn't cost much to produce, but finally the glass blowers' lungs must have given out because they eventually turned to producing floats in molds made from metal and, much later, plastic. Meanwhile, the glass-float craze spread among the world's fishermen, and soon float makers began adding their marks, embossing images such as a trademark or an artist's signature to their creations.

Japan jumped aboard the glass-float bandwagon around 1910, and of course the Chinese had to copy everyone else. During World War I, Canada's maritime armed forces used glass floats on nets intended to capture enemy submarines. But it was the Japanese who really embraced glass floats and produced millions of them ranging in size from two to twenty inches in diameter, depending on the types and sizes of nets they were intended to support. The Japanese made their floats from the glass of recycled sake bottles, most of which were clear glass, or green or brown glass. Thus the vast majority of Japanese floats found by beachcombers are of those hues. But other recycled glass was used as well, and some floats were produced in shades of aquamarine, cobalt, amethyst, yellow, orange, and, most prized of all, red, which was created by adding gold to the liquid glass. While thousands of beachcombers have over the decades hauled in millions of Japanese glass fish-net floats, mostly along the coasts of Japan, California, Oregon, Washington, British Columbia, and Alaska, Walt Pich says an equal number of glass floats remain on shore in Japan, kept in storage, and millions more are bobbing around in the North Pacific.

Around 1920, float makers began producing aluminum and cork floats, which are more durable than glass, therefore

Sake bottles came to America via the Kuroshio current.

more reliable, and which allowed fishermen to attach eye-hooks to reduce loss. By the 1940s, plastic floats were in production, and after World War II plastic became the standard for fish-net floats, boat fenders, and buoys. Few glass floats are manufactured today, though flotsamists have happy thoughts of the millions of glass balls still bobbing and weaving on the ocean currents. Don't mistake the "art objects" tossed adrift by coastal artists for the real thing.

Prince of Tides

Forks isn't one of those towns you happen upon during a leisurely Sunday drive. No, Forks lies deep in the Pacific Northwest forest, on the far side of the Olympic Peninsula, if not at the end of the world, at the end of U.S. territory. No vacationer or mountain hiker stops long in Forks, no longer than to refuel, read signs about historic logging camps, and ponder the statue of Paul Bunyan. It's not touristy, and the people of Forks like it that way. They live on the precipice of nowhere for a reason: They like solitude. From Seattle, the last real city before Forks, some 150 treacherous miles distant, the drive to Forks in good weather tops out at just around four hours. The trip involves a ferryboat ride and plenty of hairpin curve time, usually on rain-slick two-lane roads, and if it weren't for John Anderson, you would never know you'd finally reached Forks. But John Anderson's fifty-foot flotsam totem hails you on the way in.

Supported by a forty-foot crane buried somewhere inside the sculpture, the tidal totem soars above surrounding evergreens at the edge of the Anderson property. It's made completely of buoy and fender flotsam he's collected off

Pacific Northwest beaches over the past two decades. The yard surrounding his house on three sides is landscaped with flotsam. Pewter Norwegian ball floats decorate the base of a tree. A stand of crab cages resembles an early Brian Swanson sculpture. Old car fenders, rusted anchors, and massive rusted ball floats line the gravel drive. In one garden bed, nothing but washed-up pillow rock grows—not really grows, but the once-melted, now-hardened rocks seem alive and resemble dancing Dr. Seuss characters. Behind the house is a bin of more than thirty thousand plastic rice floats. The yard is only a hint, though, of the truth about John Anderson, and truth is, he owns the West Coast's, if not the nation's, largest and most extraordinary flotsam collection.

It fills a warehouse, crammed to the rafters, literally tons of flotsam: old life jackets, rusted retail signs, dozens of messages in bottles, pottery dishware, toys of wood and plastic, over a thousand glass floats including many rare varieties, shipping buoys the size of a Toyota, plastic floats of every description, Japanese wood marker stakes, gnarled driftwood monsters, old buckets, furniture frames, parts of boats and automobiles, and much more. Along one twenty-foot wall, Japanese sake bottles of every size and description occupy floor-to-ceiling shelves, neatly arranged by category. Anderson has been offered five thousand dollars for just one green glass double-doughnut float. He turned the offer down, and keeps that and other especially prized flotsam locked up inside his house about twelve inches from his peacemaker.

Hours of combing through John Anderson's flotsam collection only skims the surface; you could root around among the treasures for days and still not wrap your head around the sheer size of the collection. And it has drawn attention, not all of it welcome by the Andersons, whose home is a stone's throw from the flotsam warehouse. Anderson and his collection have been featured in *Smithsonian* magazine and on several television broadcasts, including the time he discovered remnants of a major container ship spill. Chances are that one day the flotsamist and Boy Scout troop leader,

Entrance to John Anderson's Flotsam Farm.

Prince of Tides, John Anderson, with floats he salvaged from the Pacific Ocean.

along with his wife, Debbie, who is a member of the county council, and his two sons will be obliged to move either themselves or John's collection, because the flotsam grapevine has already made Anderson's collection famous; not only other flotsamists but also the generally curious visit the Anderson place like it's a shrine.

The collection is breathtaking and marvelous, the more so because everything in the massive warehouse came out of the Pacific Ocean and was hauled up from the beach on

John Anderson's back, with the exception of a few massive pieces brought up with help, often his wife's. Among the pilgrims who visit Anderson are oceanographers who consult him frequently on what he's found and where and under what weather circumstances. Author and NOAA communications officer Robert Steelquist calls John Anderson the most interesting flotsamist he's ever known, and I agree. The man who found chow mein noodles washed up in Queets, Washington—and ate them straight from the can—is quiet, humble, and private, and now he's flummoxed, because the many visitors who show up wanting to see his flotsam collection cut into his personal time. Personally I think the Andersons should open a U.S. equivalent of the Dutch Jutters Museum, hire their two sons to run the place, and take off on a worldwide flotsam tour.

Recently the Andersons' Norwegian relatives came to visit. Maybe they saw the pewter floats encircling the tree. Maybe they recognized them.

Creative Flotsamists

One idyllic summer day, on a small island in Long Island Sound, young Jay Critchley was beachcombing the strandline when he saw a curious object on the sand. He bent over and picked it up. It had floated in on the tide, a white plastic object somewhat resembling a London bobby's whistle, yet a bit too flimsy to stop traffic.

"I liked the color and texture," says performance and fine artist Critchley, decades later recalling his first impressions of the beach whistle, "and so I started collecting them before I knew what they were."

Critchley had no clue where the beach whistles came from, or even what they were used for before they washed up on beaches in Long Island Sound. When in 1978, Critchley moved to Provincetown, Massachusetts, he was amazed at the numbers of beach whistles appearing on Cape Cod beaches. Sometime between his first beach whistle and those Critchley sees today, decades later, washing up on Provincetown beaches, somebody clued him in: Cape Cod beach whistles came from Boston's sewage system, spilling into the bay, bobbing and floating on the currents, many eventually washing up on local beaches.

There is no way to put this delicately: "Tampon applicators," explains Critchley. "And in spite of having six sisters, I didn't recognize what they were." He adds, "Until Boston fixed its sewage system, PTAs [plastic tampon applicators] were the most numerous objects that floated up on the beaches in New England. Beach cleanups would consistently rank PTAs as the number one item collected, more than bottles and cans. I began fashioning sculptures and wearable fashions."

Today, Critchley is internationally known for his art, including though not limited to his flotsam art, his outrageous theatrical installations on beaches, and grand satirical performance art aimed at, among other community causes, cleaning up Cape Cod's beaches. Critchley's flotsam art first caught my attention and rubbernecked me right into his gallery's Web site, where panning through his flotsam creations, I alternately laughed, retched, and laughed some more. In the 1980s, Critchley created a life-size Miss Tampon Liberty statue constructed entirely of tampon applicator flotsam and his most controversial flotsam sculpture, "Miss Tampon Pie." Among Critchley's inspired art works with "beach whistles," the most celebrated may be his elaborate Miss Liberty gown, created of three thousand plastic tampon dispensers. Critchley wears the gown "only on sacred occasions," for example, to the 1985 centennial celebration

Beach whistle collector, artist, and activist Jay Critchley wears his sacred robe at Provincetown (Cape Cod).
PHOTO: VINCE DEWITT/*Cape Cod Times*

for the Statue of Liberty. The full-length gown flows around his body, and when he walks or gestures, thousands of white plastic fingers twitch and flutter like a Puli's dreadlocks. On his head during sacred occasions when he has put on the robe, Critchley wears two lobster claws, the overall effect reminiscent of Native American ceremonial costumes, or a Red Lobster commercial.

In 1983, Critchley created Tampon Applicators Creative Klubs International (TACKI) to advance the collection and creative uses of PTAs. As president, he represents the organization internationally. Critchley says "there are no artificial boundaries with the oceans," and so he presides over "a vast array of collectors, folk artists, and activists. My goal was to ban the sale and manufacture of PTAs so that my collection would become valuable and allow me to retire in comfort, without relying on Social Security."

On August 29, 2000, Miss Tampon Liberty returned to the beaches for a "No with the Flow" ceremony in Provincetown, an event lamenting the opening of Boston's 9.5-mile sewage outfall pipe that jettisons some 360 million gallons per day of Boston's wastewater sewage into the waters surrounding Cape Cod and Massachusetts Bays. At the time, Critchley, decked out in beach whistle regalia, was quoted as saying, "They're still dumping, first plastic tampon applicators, now wastewater. This is a tragic and momentous occasion and a threat to our interdependence with all species. We all have a stake in the healthy and fruitful waters that sustain us here."

With the help of Clean Ocean Action of New Jersey, Critchley created a temporary beach installation called

TACKItown. He also promoted legislation at the Massachusetts House of Representatives to ban the sale and manufacture of nonbiodegradable feminine hygiene products, twice appearing in his sacred robe at hearings. Lobbyists from Playtex and Tambrands proved instrumental in defeating the bill.

Forget conch necklaces, painted driftwood beach scenes, seashell picture frames. Like Jay Critchley, John Pritchett is a visual artist whose flotsam art forms a personal bond with Poseidon. Pritchett, an award-winning editorial cartoonist for *Honolulu Weekly* and for other Hawaii newspapers, has lived in Hawaii for thirty years. Pritchett body surfs, and his favorite beach is Makapuu, at the eastern tip of Oahu. And so it's no coincidence that when John Pritchett turned to art, he would depict his favorite beach in one of his paintings. But Pritchett's seascape of Makapuu Beach, a forty-inch by thirty-inch work, was created without paint. Between December 2001 and February 2002, Pritchett collected pieces of plastic debris off Makapuu Beach and glued them onto a board to create the mosaic image. Every piece of plastic was used exactly as it was found on the beach; none were altered, except for washing off the sand.

Long considered Africa's most prolific and popular shoe, flip-flops commonly wash up on the continent's coastal beaches. Flip-flop factories in Mombasa, Kenya, produce 20 million colorful pairs a year. On Kenya's coastal islands of Lamu and Kiwayu, most folks wear pata pata, as they call flip-flops, and special cobblers make a business of repairing them until they can no longer be repaired. When monsoon rains wash trash into the ocean, the pata pata go afloat on the currents. Some stay in the ocean for years, acquiring goose-barnacle coats and hitchhikers like the swimming crab. Each year, about 12 million pairs of discarded flip-flops wash up along East Africa's coast.

When the unwearable, barnacle-coated pata pata wash up on beaches, instead of dumping the flip-flops back into trash collectors to start the cycle all over again, the islanders got creative. Today on the beaches of Lamu and Kiwayu, people harvest the colorful flip-flop flotsam, carve it into toy boats, dolphins, turtles, mobiles, and other playful objects, which are shipped to Mombasa—where the pata pata originated—and sold at market, providing a living wage for many islanders. Unlike many crafting enterprises pressed upon the poor by do-gooders, the flip-flop phenomenon was sparked and fueled by the poor. Their creativity took flotsam to another level. What constitutes a living wage is still debatable; what constitutes fashion-forward innovators is clear. About one third of today's haute couture is based on the creations of ethnic peoples, borrowed by designers and mass-producers, yet the original artists rarely reap kudos or royalties for inspiring the Paris runways and Target.

Tragic and Troubling Flotsam

For thousands of years, hurricanes and typhoons have sent oceans into tempestuous swirls while onshore winds scoop up humankind's property and debris and deposit it into the ocean. A landlubber's loss can float on ocean currents, carrying with it the memory of disaster, destruction, despair. The Boxing Day 2004 Indonesian tsunami carried offshore human remains and property, some of it washing up later along East Africa's coastline, much of it still traveling ocean currents. Similarly, Hurricane Katrina and her fellow 2005 tempests have deposited tons of property into the U.S. Gulf Coast, where, following a loop current and a strong eddy flow, treasured bits and pieces of people's lives have begun washing up along the coasts of Cuba and Mexico. A beachcomber on South Padre Island in Texas, about six weeks after Katrina, found a real estate guide from New Iberia, Louisiana. From New Iberia to South Padre Island, the guide traveled more than four hundred miles. One longtime beachcomber likened collecting disaster debris to taking pennies off a dead man's eyes.

In the spring of 1987, while photographing for a newspaper feature about a water music festival on the Washington coast, I drove one afternoon along the coast in search of the perfect beach to serve as backdrop for a cello. Eventually I found a small sandy cove surrounded by massive black boulders, the ocean foaming to shore in the background—the ideal spot. I hauled the cello down to the beach and placed it upright against some boulders, rearranging it until I was satisfied. Peering through the camera lens, the image seemed right, except for something glinting in the background off the tide line. Somewhat irritably, because a mounting storm threatened to pour rain over the borrowed cello, I trudged down to the tide line to remove the offending matter, expecting to find a soda bottle or some other innocuous object that had caught the last of the sunlight and reflected it into my image. Instead I found a pile of medical waste—dozens of hypodermic needles and syringes, and empty glass vials, their labels missing. I looked up. More vials and syringes were floating in from the ocean.

I gathered them, made the cello photograph, then photographed the medical waste for the local newspaper, and, on the way to the darkroom, dropped the medical waste off at the county health department for testing.

Three days later the results came back. I recall the editor of the *Chinook Observer* calling me into his office to deliver the news. The vials had been contaminated with, and had probably at one time contained, substances used in biological warfare.

For decades rumors have persisted that nefarious crews of ships at sea were dumping medical waste overboard. Evidence appeared on ocean beaches throughout the world. In the 1980s, around the same time I discovered the medical waste on the Washington coast, a flurry of news stories appeared in Great Britain and the United States citing mysterious medical waste, including syringes and empty vials washed up on beaches in England and Wales.

In July 1982 the U.S. Navy confirmed it normally discharged its garbage and other wastes into the Atlantic Ocean—among other sites, in the New York Bight, the area of ocean lying between Montauk Point, Long Island, and Cape May. Lieutenant Jeff Fay of the U.S. Coast Guard, stationed on Governors Island, was asked to identify the contents of a perforated galvanized canister, two and a half feet long, found bobbing on the ocean about thirty-three miles southeast of Manasquan Inlet. The canister, which was plucked out of the ocean by the crew of a sportfishing vessel, bore a red label denoting that it contained chemicals. When the Coast Guard retrieved the canister, it was, according to Lieutenant Fay, empty.

He added, "It's garbage from one of the Navy submarines. I think it's safe to say that's the way they dispose of their garbage in that area." Fay added that the Navy "apparently forgot to tie a weight" to the jettisoned canister, which, instead of sinking, bobbed across the ocean until the fishermen retrieved it.

On July 14, 1988, the *New York Times* reported used syringes and other medical waste washing up on Midland Beach on Staten Island—some, according to New York City officials, containing blood contaminated with the hepatitis B virus. The incident was just one of several on Staten Island beaches and along Rockaway Beach in Queens, where more than a dozen syringes were discovered. Despite the earlier reports of medical wastes on beaches in the United States and Great Britain, no substantial efforts addressed the potential danger until the discoveries on New York beaches. Location is everything.

Following the Staten Island incident, federal, state, and local officials convened a meeting to address the problem. While the group vowed to prosecute anyone found illegally jettisoning dangerous items into Long Island Sound, they could not agree on the origin of the medical waste. While the Rockaway Beach syringes may have been left on the beach by drug users, the contaminated vials at Midland Beach, said the City Health Commissioner, Dr. Stephen C. Joseph, were almost certainly the result of illegal dumping at sea.

But not all dangerous flotsam is jettisoned at sea: Sewage travels. Despite the effectiveness of many municipal sewage treatment systems, vast amounts of refuse from urban and small-town sewage systems escape into bays, harbors, and sounds, much of it eventually reaching the sea. While some of the floatables ride the ocean currents for hundreds, even thousands of miles, beaches in the vicinity of sewage outlets often are littered with flotsam backwash as, caught on an incoming tide, much of the escaped refuse returns to land.

Following the New York flotsam scare, Congress passed the Medical Waste Tracking Act of 1988.

Pure Garbage

The year 1987 proved a fairly trashy one for Americans, what with the Jim Bakker evangelical sex scandal and the Iran contra and Wall Street infamies, but the scandal that stank the most materialized in the form of a garbage barge named Mobro, containing thirty-two hundred tons of garbage, that traveled six thousand miles, pulled by a tugboat, searching for its ultimate destiny. Even garbage is destined, and in this case the wandering tugboat *Break of Dawn* and its famous Mobro garbage barge spurred national debate over the ultimate fate of human refuse.

The story begins with an Alabama entrepreneur named Lowell Harrelson, who experienced a personal eureka: Trash can turn a profit. Long Island had a garbage crisis. Overloaded landfills causing ground water contamination had prompted the state legislature to pass a no-landfills law to take effect in 1990, and Long Island's growing population had created a garbage emergency. Harrelson proposed barging its trash down South, where it would be dumped at cheaper landfills.

Problem was, nobody wanted a few thousand tons of New York trash. Mexico and Belize refused to take the garbage. Even Cuba refused U.S. dollars in exchange for taking New York trash. Turned away at every port where it tried to land, the garbarge Mobro—forlornly pictured on daily newscasts—continued its six-thousand-mile odyssey. Johnny Carson suggested the barge go to Iran.

Eventually, in May 1987, the Mobro anchored in Brooklyn, at Gravesend Bay, where it served as a tourist attraction until, following charges that organized crime had its foul hand in the trash, a settlement resulted in the garbage being incinerated, the ashes deposited into the Islip landfill. The Mobro's memory has entered into urban mythology, where the term "That's garbarge" is used to denounce a bad idea.

The Professionals

Last summer, at the helm of our thirty-foot 1965 Owens Flagship cabin cruiser, I revved the speed up to thirteen knots just to see the boat's wake. Puget Sound was choppy that afternoon, not many other boats out on the water, and I had a clear view ahead. The GPS told me fish were everywhere, but I'm no fisher. Suddenly I felt a thud—heard it, felt it—and fought the wheel to keep the cabin cruiser on course. Something had hit the stern, maybe disabled the rudder. I turned out of the site of the impact, circled, and saw the thing rise up on the waves. A deadhead. A huge chunk of driftwood, its mass mostly beneath the water's surface, only a small snag showing above the water. Mariners dread these stealthy floaters. Contact, especially at high speeds, can flip a boat, or at the very least damage its hull. Fortunately the Owens wasn't damaged, but it got me thinking. A sailor can be prepared for anything and still encounter the unexpected. Deadheads are one of the worst surprises on the water.

Puget Sound, being surrounded by the leftovers of clear-cut logging, is full of the stuff. Like other great bodies of water, the Sound is patrolled by a flotsam collector, a large vessel with a crane that tries to sweep the Sound free of debris. The debris is transferred to a barge and later dumped in a landfill. The flotsam collectors say they love their job, and who wouldn't during the summer months? But they're the professionals, and even in winter squalls the flotsamists are scouring the Sound for dangerous debris, deadheads being the worst. The pros have seen everything floating in Puget Sound—well, maybe not a size 40D Aubade brassiere—but they go after the dangerous stuff, and that's why the world's bays and harbors aren't more littered. Perhaps in the near future similar crews will scour the heavens, cleaning up space flotsam.

Techno-flotsam

Reports of cell phones washing onto shore add a new product to the growing list of techno-flotsam items. Other technological wonders that have washed up include balloon-borne payloads lost over the North Pacific. One payload, a research experiment for a U.S. government-funded Mars landing project, was accidentally jettisoned into the Pacific when an unexpected storm ditched the balloon and its costly hitchhiker. Only a true flotsamist would recognize the package adrift on the currents; the equipment is encased in white Styrofoam and appears to be simply a chunk of some small-time boater's ice chest. John Anderson of Forks, Washington, has recovered numerous similar payloads near his favorite beachcombing site at Queets, Washington.

The balloon launch company tracks its payloads by airplane and by monitors attached to payloads. But this time the storm prevented an aerial search, and the monitor apparently failed when it hit the ocean. The launch company, which offered a hundred-dollar reward for the balloon-borne payload, was soon contacted by a fisherman who had plucked the expensive flotsam out of a chilly ocean current on the Pacific Northwest Coast. Since the fisherman had no use for the strange package of wires and transmitters, he might have tossed it in the garbage. Instead, he made one phone call and pocketed a hundred bucks.

"You don't want this to happen," says a spokesman for the company, which prefers anonymity. "When it does, we appeal to the public's bottom line: cash reward."

On September 7, 2005, Russian cosmonaut Sergei Krikalev and his American colleague, astronaut John Phillips, aboard the joint U.S.–Russia International Space Station, jettisoned its accumulated trash—over one ton of garbage inside a Russian cargo spaceship aimed at the Pacific Ocean. The Progress 18, as the garbage ship was known, splashed down the following day at 8:13 a.m. Pacific daylight time, delivering some fascinating space flotsam into Earth's biggest flotsam collector. Included in the garbage were leftover food packages, solid waste from scientific projects, empty fuel tanks, clothing, "treated" human waste, and presumably the space station's primary oxygen generator, which had failed, causing the two astronauts to rely upon a backup oxygen supply.

The bulk of the space flotsam burned up when the garbage ship entered Earth's atmosphere. Still, some flotsam made it into the ocean. The jettisoned space flotsam generally lands in the South Pacific Ocean somewhere between New Zealand and South America.

The same day that Progress 18 splashed down, the joint space agencies launched the spaceship Progress 19 from

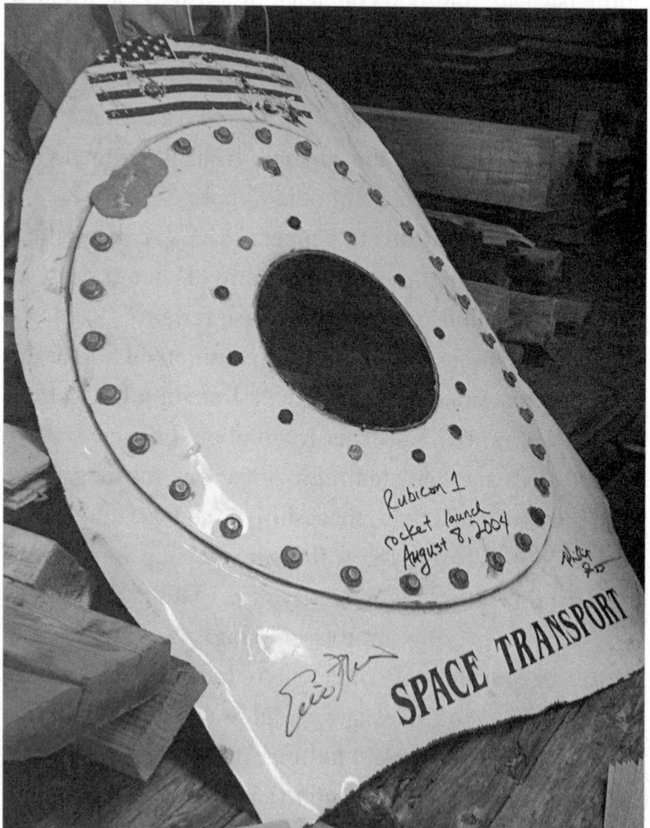

Rocket wreckage recovered by John Anderson.

Baikonur Cosmodrome in Kazakhstan, carrying 2.8 tons of food, water, technical supplies, and parts for a new oxygen generator at the space station. As with Progress 18, the Progress 19 would later become a trash receptacle, its payload aimed at the Pacific.

Ocean-borne space flotsam is a rapidly growing specialty for a certain subcult of flotsamists.

Italian Flotsam

Only a woman can truly appreciate Rome's most ubiquitous flotsam item. Ilyse Rathett, co-owner with her husband of Ritrovo, a U.S. importer of Italian food and wine, tells of the mysterious objects that for decades have washed up in abundance along the beaches at Ostia on the Mediterranean Sea. "Little sticks," she says, "thousands of them were washing up along the beaches at Ostia. For a long time, no one knew where they came from. And then someone—no doubt a woman—figured it out. Q-tips."

Why Q-tips? Rome's women wear a lot of eye makeup. Q-tips are a staple for every woman who knows how to apply and remove eye makeup. Discarded after use, the Q-tips traveled through the city's sewage system, eventually to reach the sea. The cotton tips had washed off during the product's maritime adventures, so by the time they strand, they are just little sticks.

Italian-American Flotsam

One of the more interesting bits of flotsam I have personally collected is an ID card from the order called Sons of Italy in America. "Member Name: Linda Holman. Member

#SI20328749." This bit of plastic-coated paper washed up on Alki Beach in Seattle in late 2004, after a winter storm. On the front, the image of a golden lion is inset into a double circle, with the words "Liberty, Equality, Prosperity."

Sons of Italy. I envisioned a stereotypical Tony Soprano dumping Holman's freshly killed body into the waters off Atlantic City. Somehow, Holman's corpse traveled thousands of miles, perhaps through the Panama Canal, and up along the Pacific coastline, eventually entering Puget Sound, where it beached a few dozen yards from the Coast Guard lighthouse.

A Washington, D.C., address and telephone number appear on the front of the ID, along with a fax number, but no e-mail address, indicating the card was issued before e-mail addresses were commonly added to official documents. I called the number and spoke with a receptionist who must have imagined she was speaking to a crackpot.

"Obviously," she sneered, "the member must belong to a local chapter in your area." She reeled off the telephone number of the Pacific Northwest regional director of Sons of Italy. Then she hung up. The regional rep lives in rural Washington State, and was as intrigued as I about the mysterious Ms. Holman's identity and whereabouts.

"She isn't a member of our region," the rep told me. "The Washington, D.C., office address tells me she's from the East Coast."

Touché, sneery receptionist.

I suppose Tony Soprano's boys might have transported Holman's corpse to Seattle in a Garden City garbage truck. Once here, they might have rented a boat and dumped her in Puget Sound. Why else would a decades-old Sons of Italy ID wash up on a local beach?

So, Ms. Holman, if you are still alive, and seeking your Sons of Italy ID card, please contact me, as I will be relieved to learn you were not the victim of some plot.

UFO: Unidentifiable Flotsam Obituary

Recently the pungent, decomposing carcass of what I guess was a medium-size dog, or possibly a giant opossum, washed ashore on the beach in front of my house. Gruesome, malodorous—I photographed it up close and personal—yet its Darwinian kinship sparked strong visceral empathy. Looking pathetic and foolish, the creature's remains had been

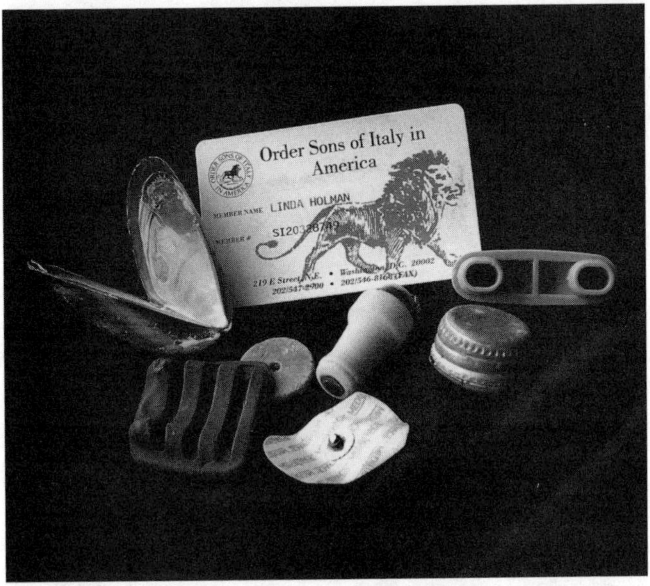

Sons of Italy ID card, medical refuse, and other tiny flotsam treasures

tossed by waves onto the rocky beach where it lay on its back, four legs akimbo, its flesh bloated and white, only a few hairs remaining on its tail and some tufts at the leg joints, its skeletal paws clean to the bone, resembling human hands, fingers clutched in agony, its headless corpse confounding immediate identification. The sight and odor triggered nausea and an odd sensation of ennui.

Odiferous Corpus Mysterious.

The carcass flotsam raised questions: What business did an opossum have riding the waves? Or a dog, for that matter? Was it surfing? Had it drowned while swimming across Puget Sound from Bainbridge Island? Had it stowed away, perhaps on one of thousands of ocean-crossing container vessels making port around Puget Sound? Was it a Japanese opossum? A Portuguese water dog? Had it been shanghaied? Or had it simply scampered out of the woods right here on Alki Beach, entered the frigid water, and suffered a heart attack? On this question I occasionally speculate, though I less often view the picture I made of its pungent, decomposing corpse. RIP, fellow air breather, and thank God for outbound tides.

Free-Range Flotsam

Annual beach clean-up events produce tons of fascinating flotsam, jetsam, and lagan. In 2005, following their annual beach sweep, the Friends of McNabs and Lawlor Islands in Nova Scotia tallied their bounty and listed it in their newsletter, *Rucksack*: "375 bags of trash, including thirty bags of recyclables. Along with the normal junk were a few unusual items such as a bingo ball (B-9), an Eastern Bakeries bread tray, a weathered plastic toy soldier, a hospital name tag, a bicycle seat with a helmet, and a TV/VCR unit, found by Clean Nova Scotia Foundation volunteers at Ives Cove.

"Plastics such as fishing gear, motor oil containers, Styrofoam, and plastic tampon applicators [oops, those beach whistles again] were among the perennial items found on the beaches again this year. The latter should be banned from all Maritime communities. . . . Volunteers . . . are tired of picking up 'beach whistles' that had been flushed down the toilets of Metro and ended up in the harbour. . . . These unnecessary products should be banned altogether or taxed heftily, to pay for their disposal."

The world's beaches accumulate tons of washed up detritus each year. Items collected during worldwide beach cleanups include bones, false teeth, clothing (socks and men's briefs seem most prolific), jewelry, watches, cell phones, television tubes, fishing gear, food and drink containers (including pottery shards from broken dishes, often originating in Asia), personal items (deodorant containers, razor blades, condoms, beach whistles), empty suitcases, plastic toys, furniture (e.g., white toilet seat), medical wastes (syringes, prescription medications, glass ampoules, illegal drugs), shopping bags (with shop names), rubber tires (often used as tugboat fenders), and empty food containers including ketchup bottles, sardine tins, and tofu wrap—all this representing the proverbial tip of the garbage heap.

Fear of Geekism

You've seen them on the beach. Usually they are large men with generous beer bellies, their Popeye forearms sunburned, their feet encased in knee-high rubber boots. Such a man wears headphones and a baseball cap, and in one hand he carries a shovel and pail while the other hand grips an object that resembles an alien's vacuum cleaner, and he sweeps it over the sand at the tide's edge as he follows the strandline in a slow, methodical gait. Occasionally he pauses and cocks his head. The alien vacuum cleaner concentrates on one spot for a while, then he sits on his haunches as he homes in on a small patch of beach.

Meanwhile, strandliners passing him gawk and sneer. Geekism personified, the beachfront bounty hunter hears not their jeers and derisive jokes; all he hears is silence interspersed with a series of intermittent tones, some high, loud and insistent, others low, sonorous and occasional. Finally the shovel comes out of the pail and the bounty hunter starts digging. Most folks are mortified for him and so they turn their heads as if he were relieving himself in the sand; he embarrasses the heck out of them, but they can't say why.

It's just that he looks so . . . geeky.

When I ordered my Bounty Hunter Scout Metal Detector, I had visions of my husband, or a friend, accompanying me to the beach. Fat chance. Apparently I know no one willing to risk his or her reputation being seen on the beach with a metal detector person. I called everyone I knew, even my own siblings, to no avail. You'd think I was planning to streak the beach—no one agreed to join me. When I say no one, this includes my editor, who, when I told him I needed someone to come along and photograph me actually operating a metal detector, said, "I don't blame your husband for turning you down. I wouldn't do it either."

And so my Bounty Hunter Scout arrived. On the next low tide, I disguised myself beneath bulky clothing, hat, and sunglasses and hit the beach, Scout in hand. Instantly my neighbor's two sons, ages five and seven, recognized me.

"Oh cool," said Jack. "Let's go treasure hunting." I should have thought of this before—kids have no inhibitions, and they're willing to give geeks a chance.

Headphones in place, the Scout's dials turned up full blast, I moved along as the instruction manual suggests, slowly sweeping the alien vacuum cleaner across the beach. The boys, Jack and Max, scampered beside me. Nothing happened for a while, and then, "Screeeech!" I turned down the Scout's volume and began digging where the metal detector had indicated buried treasure. You've got to be patient; kids aren't known for long attention spans. After watching me dig for ten minutes, Jack and Max, to my extreme mortification, outed me to their approaching father and, adding insult to injury, deserted me to go off with their dad for hot chocolates. Who could blame them?

Alone now, digging in the sand as numerous strandliners walked by and gawked, I became acutely aware of a woman watching me. I placed my cell phone to my ear and pretended to be having a conversation, as if this made me look more normal. I could literally feel the strandliners' sneers and, turning my back, kept digging. Ah, here it is. Yes, ladies and gentlemen, the metal object is . . . what? I believe it's a rusted arrowhead from the Iron Age. That or a section of rusted drainpipe. Yep, drainpipe metal. But the Scout worked, and I was motivated. Three hours, forty sneers, and fifteen holes later, I came home with the faux Iron Age arrowhead, three rusty nails, a rusted Corona beer bottle cap, an electrical fitting of some sort, a rusted wheel, perhaps off a chair or stool, a rusted ring missing its diamond, and an unidentifiable blob of deeply pitted, rusty metal.

Originally I had intended to use the metal detector once and then ditch it. Poring over my meager first-dig treasures, though, I asked myself, "Was it worth all the scorn and ridicule?" Surprisingly, the answer was, "Yes, indeed." I hadn't had this much fun since I was five years old trying to dig a hole in the beach from here to China. Now whenever I go metal detecting for beach treasure, I still go disguised, and with a cell phone to symbolize conformity, but in my chest burns a smug satisfaction, for one day I will surely strike gold.

The epiphany shouldn't have surprised me. I suppose my psychiatrist had known it all along. I believe it hit me around the third or fourth time out with the metal detector. I was wearing my wet suit, wading knee-deep along the tide line, the Scout hovering just above the water's surface. If anyone was sneering, I didn't hear it. I was wearing a set of Bose headphones, listening for the low beep that signaled gold. Instead, I heard a voice, maybe my own, recite as if from a news banner, "Oddball Discovers True Calling as Flotsamist."

He's everywhere. Nixon in Japan.

There it was. The answer to Curious Question number two: Was I a mere beachcomber or a genuine flotsamist? I couldn't wait to tell the world. My husband. My psychiatrist. Like a child taking a first step, or a Scientologist reaching Full Clear, as if a rogue wave's fetch had swept me up in its cleansing maw and deposited a reborn me on the tide line, I wanted to shout above the roaring sea, "I am a true blue

Metal detecting is a popular hobby around the world. Germany makes the most sophisticated metal detectors for beachcombers.

flotsamist." The answer to Curious Question number one, regarding the mysterious floating stone, would soon follow, or so I thought.

Strandings

Strandings—what an interesting word. A beach or shoreline is also called a strand, and so it makes sense to say a whale washed ashore has been stranded. Boats are stranded, people are stranded, anything washing ashore is stranded. Flotsam discovered on beaches has been stranded by an ebbing tide. A tax some coastal communities levy on fish or other washed-up commodities is known as strandage. According to international maritime law, whoever finds flotsam on the high seas or washed up on beaches may take rightful ownership unless the flotsam is clearly marked as property of a government, or unless its previous owner can prove rightful possession.

I'm wondering if I should try to locate that Japanese boy whose pink plastic airplane propeller I found washed up. And then it dawns. As gently as my forehead clears of lines, a slow chill travels from a sprung brain gate down along my spine, where it comes to rest in the pit of my stomach.

Pit. Peach pit on steroids. I dive for Ed Perry's book on drifting seedpods. I thumb through it, then skim, then go page by page. Not there. Ah well, sometimes you just have to let go of your flotsam dreams. Eventually I removed the pictures of the strange floating stone from my bathroom mirror, from my car's dash, from my desk. But I never forgot it, and to honor its place in my flotsaming life, I made a screensaver of the picture and moved on as flotsamists will, to seeking new treasure on the tides.

One day near summer's end I plucked a book on Hawaiian plants from the shelf and leafed through it. I'd been considering paring down my library and this seemed a good place to start. I mean, come on, why did I need a book on Hawaiian garden plants? And there it was.

Stared me right in the face. My beach stone. An exact replica. The plant's name: Hawaiian country apricot. A cold thrill embraced me. Could it be? Was it possible? Well, it had happened. I had proof, of sorts. A photograph of the floating stone on beach gravel—not Hawaiian sand. But would anyone believe me? Would they think I'd gone to the Hawaiian countryside and picked up a country apricot seed, brought it home, and placed it on my own beach? I know someone who travels the world's beaches and places non-native stones everywhere, just for the fun of confusing folks. Had someone done this to me? No, indeed. The stone I found had washed up.

I'll keep on searching for it, mentally castigating myself for leaving the flotsam treasure of a lifetime on the beach

See why this Hawaiian apricot seed was mistaken for a floating rock?

where I'd found it, after it had traveled, at the very least, according to the calculations of the late Amos Wood, from Hawaii, for seven years via the Kuroshio Current some eleven thousand miles in circular Pacific Ocean currents, before washing up on Alki Beach, where I had fondled and wondered over it before recklessly tossing it into the water, where the next tide carried it back out to sea, or with luck, on a circuitous journey around Puget Sound. And so all you fellow flotsamists and beachcombers, please take note: Dibs, I saw it first, and I've got the picture to prove it. If that isn't enough, I'll swap out my metal detector for that marvelous floating stone.

So my psychiatrist said, "Now maybe you'll settle down and face your real demons. Get over this."

"What? Just because I've finally discovered a globally thriving social group into which I may comfortably fit? Just because I've figured out that I'm as geeky and obsessed as other flotsamists? Hey, doc, I've dumped my demons. Traded them in for a house full of flotsam. Ever since I discovered the origin of the floating stone, I'm cured."

Imagine a psychiatrist rolling his eyes. He said, "You think this, um, Hawaiian country apricot seed is worth anything?"

"Hey, I'm a flotsamist," I said. "I'd pay at least twice what I pay you every year for that lost treasure."

"So you'd know this Hawaiian apricot seed when you see it?"

"Oops," I said, "time's up. And according to my tide tables, I have twenty minutes to reach the beach before the tide turns. After that storm last night, I can't wait to see what washes up."

"Hold on," said the shrink. "I'll get my coat."

Acknowledgments

The following individuals provided valuable information, contacts, and anecdotes, for which I am most grateful. None, however, is responsible for any errors and/or omissions inadvertently perpetrated in this book; for these, I shoulder full responsibility and invite all naysayers to bend my ear if you can find me on the beach.

Thanks to: Professor Tadashi Ishii, Fukuoka, Japan; Captain Charles Moore and crew of the research vessel *Alguita;* Robert Steelquist, education and outreach coordinator, Olympic Coast National Marine Sanctuary, NOAA; Neil and Kathleen Robbins, Newport, Oregon; oceanographers W. James Ingraham, PhD, and Curtis C. Ebbesmeyer, PhD, Washington state; reporter Luann Swanson, *Tillamook Headlight-Herald,* Tillamook, Oregon; Bill DeSousa and the Cape Cod Chambers of Commerce; artist Joan Lederman, Woodshole, Massachusetts; R. Hoarau, Mt. Kenya Safari Club, Nanyuki, Kenya; Datom Margishvili, Tblisi, Georgia; Jukka Petaja and Alpo Suhonen, Helsinki, Finland; Gregory Smith, San Clemente, California; Wakako Otake, Tokyo, Japan, sister, reader, and keen critic, Elizabeth M. Speten, the forgotten lascars of the world; copy editor Don Graydon—a miracle worker; Dana Youlin, Kurt Stephan, Liza Brice-Dahmen, Austin Walters, and all the Sasquatch staff; and, most importantly, special thanks to editor Gary Luke, who, in spite of refusing to be seen on a public beach with me and my metal detector, nevertheless smiled when I first mentioned flotsam, and kept smiling as he yanked the proverbial pencil from my graspy fingers.

I am especially grateful to my husband and fellow writer, G. M. Ford, my lighthouse and beacon.

Suggested Reading: Books on Flotsam and Related Subjects

"A Rare Species of Ambergris." *Parfumes Cosmetiques Actualites* 175 (Feb/Mar 2004), p. 28.

Awano, K., S. Ishizanki, O. Takazawa, and T. Kithara T. "Analysis of Ambergris Tincture." *Flav & Frag Journal*, 20 (2005), pp. 18–21.

Burfield, Tony. *Natural Aromatic Materials—Odours and Origins*. Tampa: Atlantic Institute of Aromatherapy, 2000.

The Columbia Encyclopedia. New York: Columbia University Press, 2005.

Dennis, John V., and Charles R. Gunn. *World Guide to Tropical Drift Seed and Fruit*. New York: Quadrangle/The New York Times Book Co., 1976.

DeWire, Elinor. "Mooncussers." *The Compass* 3 (1989).

Ebbesmeyer, Curtis C. *Beachcombers Alert!* Seattle: Published by Curtis C. Ebbesmeyer, quarterly.

Felter, Harvey Wickes, and John Lloyd. *King's American Dispensatory*. Cincinnati: Ohio Valley Co., 1898.

Ford, Peter. "Drifting Rubber Duckies Chart Oceans of Plastic." *Christian Science Monitor*, July 31, 2003.

Green, Alan A. *Jottings From a Cruise*. Seattle: Kelly, 1944.

Guppy, H. B. (1917) *Plants, Seeds and Currents in the West Indies and the Azores*. London: Williams and Norgate, 1917.

Ishii, Tadashi. *Encyclopedia of Flotsam:* Encyclopedia-shinpen hyouchakubutsu gitenn. 2nd edition. Tokyo: Kaichousha Co., 2002.

Jebens, Holger, ed. *Cargo, Cult & Culture Critique*. Honolulu: University of Hawai'i Press, 2004.

Johnson, Captain Charles. *A General History of the Robberies & Murders of the Most Notorious Pirates*. New York: The Lyons Press, 1998.

Kinder, Gary. *Ship of Gold, In the Deep Blue Sea*. New York: Vintage, 1998.

Le Galliene, R. *The Romance of Perfume*. New York: Richard Hudnut, 1928.

Menard, Wilmond. "Neptune's Sea-Mail Service." *Sea Frontiers* 26:6 (Nov/Dec 1980).

Moody, Skye [Kathy Kahn]. *Fruits of Our Labor*. New York: G.P. Putnam's Sons, 1982.

Moore, Charles. "Trashed: Across the Pacific Ocean, Plastics, Plastics, Everywhere." *Natural History Magazine*, vol. 112, no. 9, Nov 2003.

Moore, Ellen J. *Fossil Shells From Western Oregon*. Corvallis, OR: Chimtimini Press, 2000.

Morgan, Curtis. "LEGOs and Other Floating Flotsam." *Miami Herald*, May 17, 1998.

Mugurevics, E. *Viking Age and Medieval Finds of East Baltic Amber in Latvia and the Neighbouring Countries (9th–16th Century)*. Riga: Amber in Archaeology, 2003.

Out in the Pacific, Plastic is Getting Drastic (ship log of Capt. Charles Moore, aboard oceanographic research vessel Alguita, October 22, 2002). *www.mindfully.org/Plastic/*.

Perry, Ed. and John V. Dennis. *Sea-Beans from the Tropics*. Malabar, FL: Krieger, 2003.

Petitdidier J.P. "Fixeurs Animaux: L'ambre, le castoreum, la civetter, le musc." *Parfums, Cosmétiques, Arômes* 90 (Dec 1989/Jan 1990), pp. 79–82.

Pich, Walt. *Beachcombers Guide to the Northwest: California to Alaska*. Ocean Shores, WA: 1997.

Pich, Walt. *Glass Ball*. Ocean Shores, WA, 2004.

Steinem, Gloria. *The Beach Book*. New York: The Viking Press, 1963.

Synthetic Sea: Plastics in the Open Ocean (a film produced by MacDonald Productions/Mindfully.org). Long Beach, CA: Algalita Marine Research Foundation, CA, 2001.

Tennessen, J.N., and A.O. Johnsen. *The History of Modern Whaling*. Berkeley: University of California Press, 1982.

Webber, Bert and Maggie Webber. *I'd Rather be Beachcombing*. Medford, OR: Webb Research Group, 1993.

Wood, Amos L. *Beachcombing for Japanese Glass Floats*. Portland, OR: Binfords & Mort, Portland, 1967.

Wood, Amos L. *Beachcombing the Pacific*. West Chester, PA: Schiffer Publishing Ltd., 1987.

Web Resources

Algalita Foundation: *www.algalita.org*

American Meteorological Society: *www.ams.allenpress.com*

CargoLaw, Law Offices of Countryman & McDaniel: *www.cargolaw.com*

DiscoverSea Shipwreck Museum of Fenwick Island, Delaware: *www.discoversea.com*

Dr. C's Remarkable Ocean World: *www.oceansonline.com*

The Infamous Exploding Whale: *www.perp.com*

International Colored Gemstone Association: *www.Gemstone.org*

International Movie Database (IMDb): *www.imdb.com*

International Sea-Bean Symposium: *www.seabean.com*

Irish Seaweeds: www.irishseaweeds.com

Tadashi Ishii's official Web site of "What the Ocean Brought": *www.polepoleto.com/boboforest/home/t-isii/1.html*

Kellyco Metal Detector Superstore: *www.kellycodetectors.com*

Maritime and Jutters Museum of Texel, The Netherlands: *www.texelsmaritiem.nl*

Captain Charles Moore: *www.alguita.com*

National Oceanic & Atmospheric Administration: *www.noaa.gov*

Rice University Forum: *www.ruf.rice.edu*

United States Office of Naval Research: *www.onr.navy.mil*

About the Author

Writer, photographer, and former East Africa bush guide, **Skye Moody** is the author of two previous books of nonfiction, and seven books of fiction focusing on environmental issues. As a journalist she has covered Chinese and Ukranian coal mining, reindeer herding in Siberia, textile mills and farming in Uzbekistan, river pollution in the Republic of Georgia, and the effects of acid rain in the Arctic Circle, winning critical acclaim for her books of nonfiction, *Hillbilly Women* and *Fruits of Our Labor.* She is a member of PEN American Center.